FLOOD PROBLEM AND MANAGEMENT IN SOUTH ASIA

Flood Problem and Management in South Asia

Edited by

M. MONIRUL QADER MIRZA
Adaptation and Impacts Research Group (AIRG), Institute for Environmental Studies (IES), University of Toronto, Canada

AJAYA DIXIT
Nepal Water Conservation Foundation (NWCF), Kathmandu, Nepal

AINUN NISHAT
IUCN – World Conservation Union, Dhaka, Bangladesh

Reprinted from *Natural Hazards*,
Volume 28 No. 1, 2003

KLUWER ACADEMIC PUBLISHERS
DORDRECHT/BOSTON/LONDON

A C.I.P. Catalogue record for this book is available from the Library of Congress.

ISBN 1-4020-1068-0

Published by Kluwer Academic Publishers,
P.O. Box 989, 3300 AZ Dordrecht, The Netherlands.

Sold and distributed in the U.S.A. and Canada
by Kluwer Academic Publishers,
101 Philip Drive, Norwell, MA 02061, U.S.A.

In all other countries, sold and distributed
by Kluwer Academic Publishers,
P.O. Box 989, 3300 AZ Dordrecht, The Netherlands.

Printed on acid-free paper

Printed in the Netherlands

Table of Contents

Natural Hazards **28:** vii–ix, 2003.

Preface

South Asia faces flood disaster on a regular basis. This problem crosses international borders and as the regional hydrology is affected by climatic change, the nature of the problem has become complex. Past approaches focused on altering of stock and flow of water using embankments and reservoirs, but are insufficient. The structural solutions are inadequate to address the problems because they require collaboration (such as, regulation and joint planning/implementation of water control structures, well maintained sluice gates and embankments) that are difficult to achieve both within and between countries.

The structural solutions to be effective require major restructuring of both legal systems and institutions responsible for management. However, little social capacity exists at present to bring about such transformations in South Asia. In addition, the prevailing social and political contexts make them difficult to achieve. As a result, integrated management approaches to flooding advocated by hydrologists and water resource specialists are inadequate to address the challenges that the countries of South Asia face.

Given the above limitations on the ability to control water to avoid flood disasters, strategies that address emerging challenges will be needed. Flood disaster has differential impact on individuals, households and communities. People cope in different ways. Those who have the capacity after being hit by a disaster emerge faster while those without such capacity sink deeper into the spiral of impoverishment. Coping strategies include actions such as migration from flood-affected areas, flood forecasting, flood insurance of animals and crops, food stockpiling, providing emergency health services and building flood shelters. They have, however, not been woven systematically into the approach to achieve security from flooding.

If the approaches build on coping strategies and seek to identify new ones, they could address the social impacts of flood problems effectively at a lower social, economic and environmental cost than approaches that attempt to manage or control the resource base itself. Furthermore, responses essential to such strategies would encourage collaboration among and between countries of South Asia that will be rooted on the nature on the problem rather than a grand narrative centered only on specific control/mitigation method.

This special issue of *Natural Hazards* looks at the flood scene in South Asia by focusing on both the hazard and the vulnerability aspects including their assimilation. Its conceptual foundations suggest the need for adaptive approaches to management of risks related to floods and to identify specific strategies that could help people overcome the social, economic and environmental vulnerabilities emerging from flooding. These issues are addressed by the authors who focus on hazard and specific flood events including the social, economic and institutional processes.

The first article by O. N. Dhar and Shobha Nandargi focuses on hydrometeorological aspects of floods in India. The paper suggests that the ground conditions are quite favourable in generating high run-off because of the antecedent wet conditions that the monsoon rains create. On the basis of a detailed discussion of floods that span fifteen years in different river systems the paper shows that the problem is confined to the states located in the Indo-Gangetic plains, northeast India and occasionally in the rivers of Central India.

M. Monirul Qader Mirza analyses three extreme flood events in Bangladesh. These were in 1987, 1988 and 1998. These floods differed in terms of magnitude, extent, depth and duration. The external and internal hydro-meteorological dynamics were also different.

Vishwas S. Kale analyses the geomorphic effects of monsoon floods in Indian rivers. The geomorphic effects of floods are evident only in certain areas – the Himalayas, the Thar desert, the Indus and Brahmaputra plains. The peninsular rivers are, by and large, stable and the geomorphic effects of floods are modest. A synthesis of the various case studies available from the Indian region indicates that often the absolute magnitude of a flood is not as important with respect to the geomorphic effects as the flow stress and competence.

I. M. Faisal, M. R. Kabir and A. Nishat analyse the impact of the 1998 flood on Dhaka City. Water entered into the protected part of the city through poorly maintained infrastructure. Lack of coordination between the agencies responsible for flood protection and drainage of the city exacerbated the problem. Non-structural measures involving retention ponds, maintaining of drains, introducing land zoning and flood proofing in the eastern part of the city can help mitigate flooding. Institutional bottlenecks need to be overcome.

Harun Rasid and Wolfgang Haider present floodplain residents' preference for flood water level management options in flood control projects in Bangladesh. For this purpose, a maximum difference conjoint (MDC) model was employed as a part of a questionnaire inside the Compartmentalization Pilot Project (CPP) – Tangail. The results of the study indicated that the respondents had a clear preference for preventing flooding of their homes and courtyards and rice fields. They also expressed a strong concern about malfunctioning of sluices and to a lesser extent about the changes in the fish habitat.

P. K. Mohapatra and R. D. Singh addresses flood problems in India, regional variability of the problem, present status of the ongoing management measures, their effectiveness and future needs in flood management. Some special problems related to floods like dam break flow, and water logging are also mentioned.

Richard Kattelmann analyses mitigation of at glacial lake hazard in the Himalayas. Glacial lake outburst floods have become an active topic of discussion within the development community focused on Nepal. The first attempt within Nepal to reduce hazard of one lake artificially by lowering its water level was partially completed in June 2000. Beginning work on other hazardous lakes will require decisions about investment from the international aid community.

Ajaya Dixit provides a holistic picture of flood disaster in Nepal. The paper suggests that vulnerability of people in areas prone to flooding must be addressed by enhancing their resilience capacity. The approach to flood mitigation must be pluralistic and give space to different management styles with varying obligations at varying scales.

A Bangladeshi perspective on regional cooperation in flood management in the Ganges–Brahmaputra–Meghna (GBM) catchments are presented by Q. K. Ahmad and Ahsan Uddin Ahmed. The existing flood forecasting and warning capacity of Bangladesh could be made effective if real-time data could be obtained from upstream areas of the Ganges, Brahmaputra and Meghna (GBM) catchments. To that end, Bangladesh needs to foster an effective regional cooperation mechanism with the other GBM regional countries.

Sujata Gupta, Akram Javed, and Divya Datt present the economics of flood protection in India. Forty million hectares of land in India is prone to floods. Though 48% of the vulnerable area has been provided with reasonable protection, floods losses continue to increase. This paper presents a simple regression exercise for three states highly vulnerable to flood, and argues that flood protection measures have been inadequate in controlling losses and reducing vulnerability.

Collectively, the papers in the volume suggest that diverse causes determine severity of the events. At the same time, the flood has differential impact on the societies hit by floods.

We would like to thank many individuals in Asia, Europe, Australia and New Zealand, North and South America who offered their valuable time and expertise, and through reviewing the manuscripts ensured the high technical standard of this special issue. We specially extend our thanks to Dr T. S. Murty, Editor (North America) of *Natural Hazards* for his outstanding interest on this topic and pioneering this volume of the Special Issue. We wish to thank Roger Street, Abdel Maarouf and David Etkin at the Adaptation and Impacts Research Group (AIRG), Environment Canada and Dr D. K. Barua, Rescan Environmental Services Inc., Vancouver for their cooperation.

M. MONIRUL QADER MIRZA
AJAYA DIXIT
AINUN NISHAT

Natural Hazards **28:** 1–33, 2003.

Hydrometeorological Aspects of Floods in India

O. N. DHAR and SHOBHA NANDARGI
Indian Institute of Tropical Meteorology, Pune - 411 008, India

(Received: 27 February 2001; accepted: 18 July 2002)

Abstract. The Indian sub-continent being located in the heart of the summer monsoon belt, receives in most parts more than 75% of its annual rainfall during the four monsoon months of June to September. As the bulk of summer monsoon rainfall occurs within a period of four months, naturally majority of floods occur in Indian rivers during this season only. The ground conditions also help in generating high percentage of run-off because of the antecedent wet conditions caused by rainy spells occurring within the monsoon period itself. Besides mentioning different weather systems, which cause heavy rainfall and consequent floods, a detailed discussion of 15 years' floods in different river systems has also been given in the article. This study has shown that the flood problem in India is mostly confined to the states located in the Indo-Gangetic plains, northeast India and occasionally in the rivers of Central India.

Key words: Hydrometeorology, floods, cyclonic storms, seasonal monsoon trough, break monsoon situations and rainstorm zones

1. Introduction

India being an agricultural country, the prosperity of its agriculture largely depends upon the rainfall received during its summer monsoon. Considering the beneficial aspects of summer monsoon rainfall to its agriculture, a British Member of the then Viceroy's Executive Council said that India's Budget is a gamble on its summer monsoon. It is well known that summer monsoon rainfall is not uniformly distributed over the Indian region and there are large variations in its space-time distribution over different regions. Some parts get heavy to very heavy rainfall while others get little rainfall. In the long history of about 130 years' (1871–2000) of rainfall records of this country, it was observed that highest rainfall over the country was of the order of 22% above the normal annual rainfall of 117 cm in the year 1961 and it was below normal by about 30% in the year 1877 (Sikka, 2000). In the remaining years the mean annual rainfall ranged between these two extreme limits. This unequal distribution of monsoon rainfall is one of the major causes of floods in the Indian rivers.

Of all the natural disastrous events, floods by far are the most hazardous, frequent and widespread events throughout the world. According to WMO (1994), on the basis of data of major disasters that occurred around the world during the period 1963–1992, it has been found that floods cause the maximum damage as shown below:

Type of hazard around the world	*Damage caused by natural calamities*
1. Floods	32%
2. Tropical cyclones	30%
3. Droughts	22%
4. Earthquakes	10%
5. Other disasters	6%

It is also found that maximum number of deaths (i.e., 26%) are caused by floods while only 19% are caused by tropical cyclones. Natural calamities like earthquakes and droughts cause 13% and 3% of deaths respectively (WMO, 1994).

In view of the above, an attempt has been made in this article to give a brief resume of floods that have occurred during the 15-year period (1986–2000) of the summer monsoons. Before that is done, the main hydrometeorological aspects which cause floods in the Indian river systems, year after year during the summer monsoon months are discussed.

2. Mean Summer Monsoon and Annual Rainfall over India

Indian area is of sub-continental size. Its area is about 3,287,263 km^2 and it is located between Long. 68°07′ and 97°25′ East and Lat. 8°09′ and 37°06′ North. In the north, it is bound by the Great Himalayan range which runs from east to west in the form of convex arc whose length is about 2400 km. Its width varies from about 250 km to about 400 km and it contains some of the highest mountain peaks of the world like Mt. Everest (8848 m), Mt. Kanchenjunga (8611 m), Dhaulagiri (8172 m), Annapurna (8078 m), Nanga Parbat (8126 m), etc. Figure 1 shows the broad orographic features of the Indian sub-continent. Some 22 major rivers of the Indian subcontinent drain the Himalayas from east to west and bring their waters into the northern plains of India where they join the Ganga and the Indus rivers of the sub-continent. The source of all their waters in these rivers is contributed by the melting of glaciers and winter snows of the Himalayas besides the summer monsoon rainfall.

Long period (1891–1970) average annual rainfall over the contiguous Indian area is about 117 cm (Dhar *et al.*, 1981a) which is the highest for a country of comparable size anywhere in the world. It is however seen that the summer monsoon rainfall is highly variable both in time and space. Almost 75% of the long period average annual rainfall comes down in the four monsoon months of June to September. The heaviest rains of the order of 200–400 cm or even more occur over northeast India and along the Western Ghats situated along the west coast of the peninsular India. On the other hand some regions in the extreme western part of the country, such as western Rajasthan, receive average annual rainfall which is of the order of about 15 cm or even less. By and large, the annual average rainfall over the

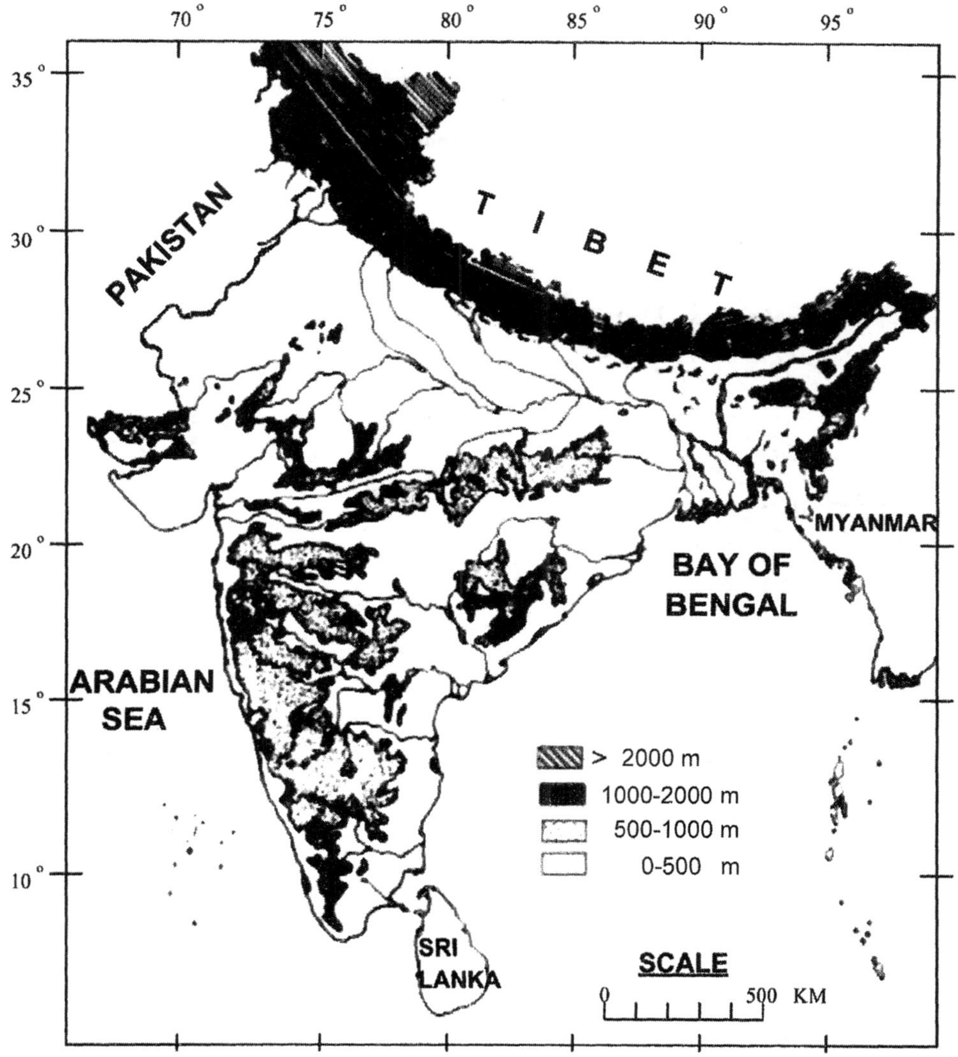

Figure 1. Ororaphy of the Indian subcontinent showing regions higher than 500 m, 1000 m, and 2000 m.

northern Indo-Gangetic plains running parallel to the foot-hills of the Himalayas, varies from about 150 cm in the east to 50 cm in the west. Over the central parts of the India and northern half of the peninsular India, it varies from 150 cm in the eastern half to about 50 cm on the lee side of the Western Ghats. In the southern half of the Indian peninsula average annual rainfall varies from 100 cm to 75 cm as we go from east to west. Figure 2 shows the distribution of average annual rainfall over India.

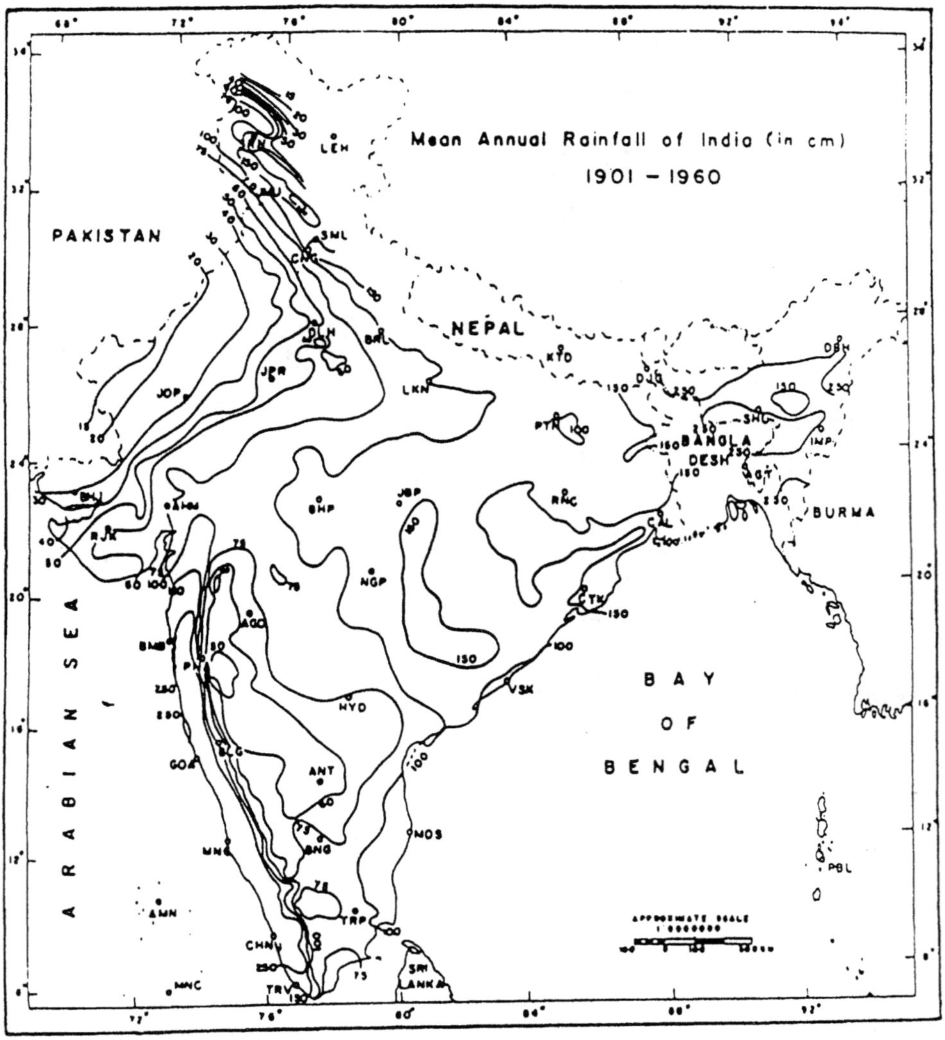

Figure 2. Mean annual rainfall of India (1901–1960) in cm.

3. River Systems of India

There are about 120 major and medium rivers in India which criss-cross the Indian area from north to south and east to west. The number of minor rivers that flow over different parts of the country is legion. There are about 340 gauge/discharge (G/D) sites located along these major and medium rivers, whose river water levels are regularly measured especially during the monsoon months. The data of floods given in this article pertain to some of these sites. The classification of Indian rivers mostly followed in the context of floods, is as given by Framji and Garg (1976), which divides Indian river systems in four groups on the basis of meteorological, geological and topographical conditions. These river groups are as follows and these are also shown in Figure 3 :

(a) Brahmaputra river system,
(b) Ganga river system,
(c) Northwest river system and
(d) Central India and Deccan river system.

4. Definition of Floods

Flood has been defined by various workers/organizations in various ways. Following are some of the definitions given in the literature on floods:

Chow (1956) described flood as a relatively high flow which overtakes the natural channel provided for the runoff. According to Ward (1978), flood is a body of water which rises to overflow the land which is not normally submerged. According to World Meteorological Organization (WMO, 1974), a flood is a rise usually brief, in the water level in a river to a peak, from which the water level recedes at a slower rate.

In India, in the common parlance, a river is said to be in flood when its water level crosses the Danger Level (D.L.) at that particular site (Dhar and Nandargi, 1998). Danger level for a site is usually one meter above the warning level (W.L.). This level when attained by a river at a site will initiate issuing flood warnings as flood damage begins at this level (Kulshrestha, 2000). When a flood crosses the D.L. by 1 meters or more, it is termed as '**major flood**' at that site; when it is 5 meters or more it is called '**severe flood**' and floods 10 meters or above the D.L. are called '**devastating floods**'.

5. Basic Hydrometeorological Weather Systems which Cause Floods in India

As mentioned earlier, intense rainfall is the main cause responsible for causing floods in India. The ground condition, on which intense rain falls, plays an important role in generating flood run-off from rainfall. In the following paragraphs major flood producing synoptic weather situations are described in detail.

5.1. CYCLONIC DISTURBANCES

During summer monsoon months (June to September) floods in Indian rivers are caused by rainfall that occurs in association with the movement of monsoon depressions and cyclonic storms through the country. These disturbances occur very frequently during the summer monsoon period of June to September and mostly originate from the Bay of Bengal and very rarely from the Arabian Sea. Figure 4 shows the spatial distribution of depressions and cyclonic storms over India during the period of 1891 to 1980 (Dhar *et al.*, 1984a).

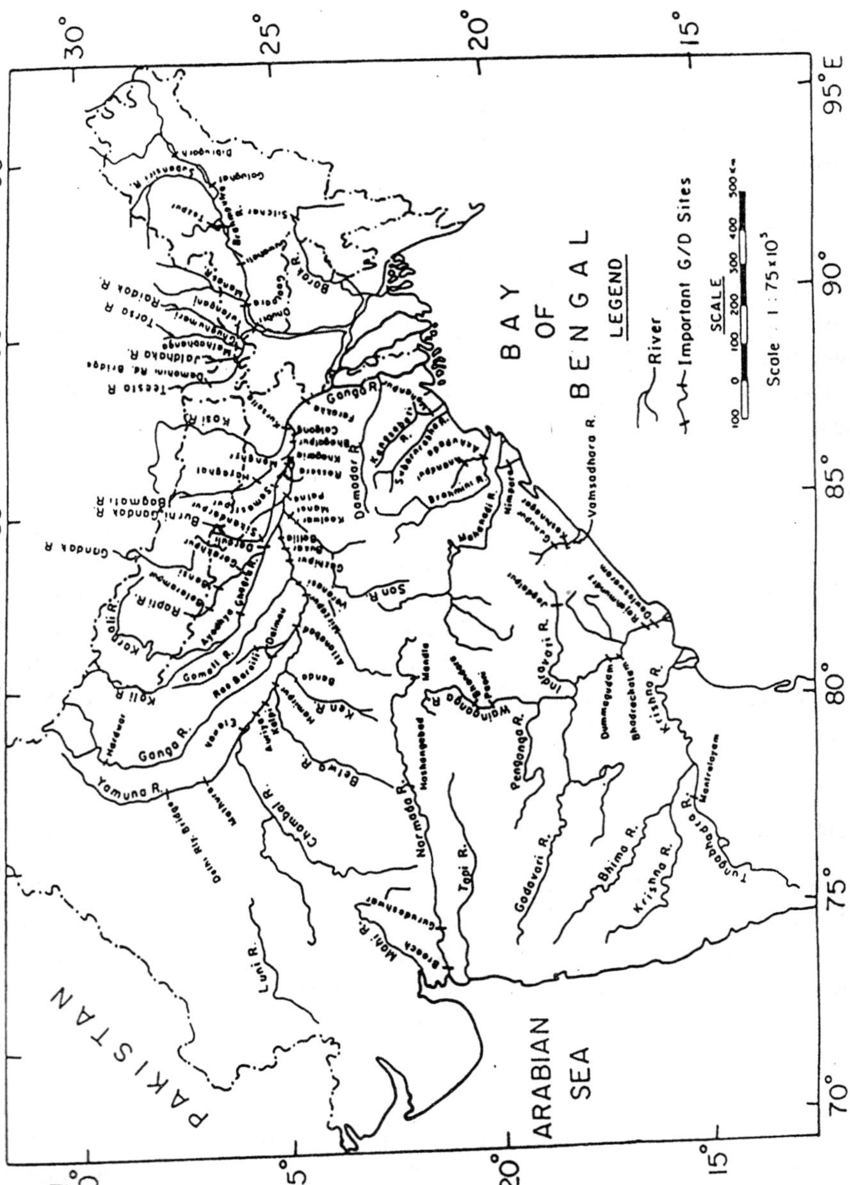

Figure 3. Map of India showing major river systems and gauge/discharge (G/D) sites on them.

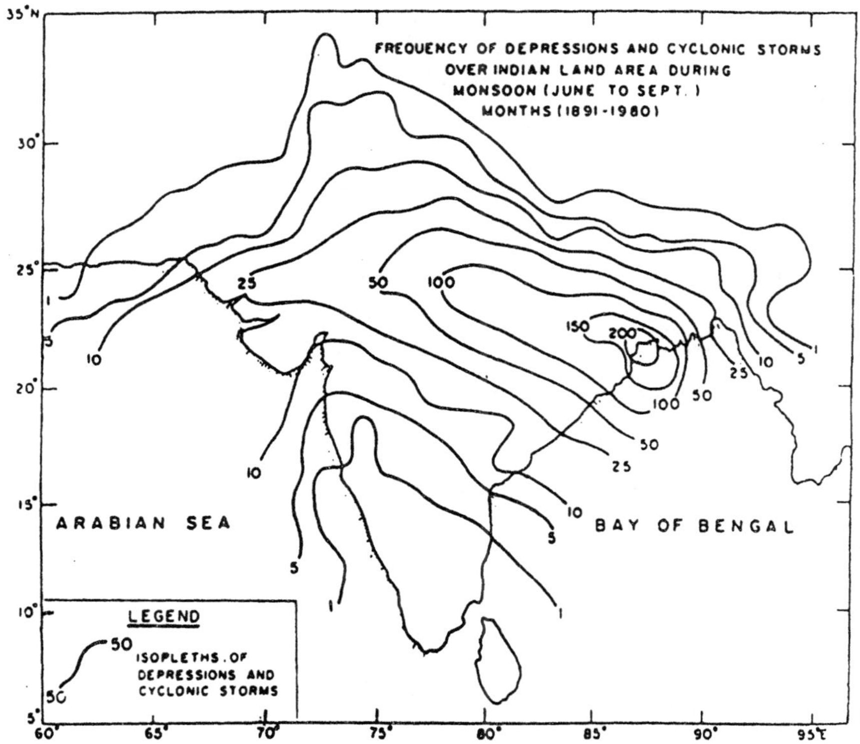

Figure 4. Frequency of depressions and cyclonic storms over India during 1891–1980.

These cyclonic disturbances during the monsoon months mostly form in the head Bay of Bengal and travel in a northwesterly to westerly direction through the Indo-Gangetic plains and its neighbourhood after crossing the coast. Sometimes, after moving far into the interior of the country, these disturbances recurve and move towards north or northeast and break over the foothills of the Himalayas under the influence of westerly troughs or western disturbances (WDs) moving from west to east, at higher latitudes. Associated with these cyclonic disturbances, both the Arabian Sea and Bay of Bengal branches of the summer monsoon currents strengthen considerably and cause heavy to very heavy rainfall over the region these disturbances move through. It is, however, observed that in summer monsoon season, heavy rainfall generally occurs in the southwestern sector of these disturbances (Pisharoty and Asnani, 1957). Heavy rainfall also occurs in the northeastern sector when they recurve from their westerly to northwesterly track towards the north or northeast.

Large floods on most of the Indian rivers are a direct result of intense cyclonic storms and depressions. The flood-generating rainstorms are confined to two major zones (Nandargi, 1996). The first zone, i.e., 'Zone A' is located over the basin region of Ganga and its tributaries and the plains of Punjab and the second zone, viz. 'Zone B' which includes the central India and northern half of the Indian peninsula

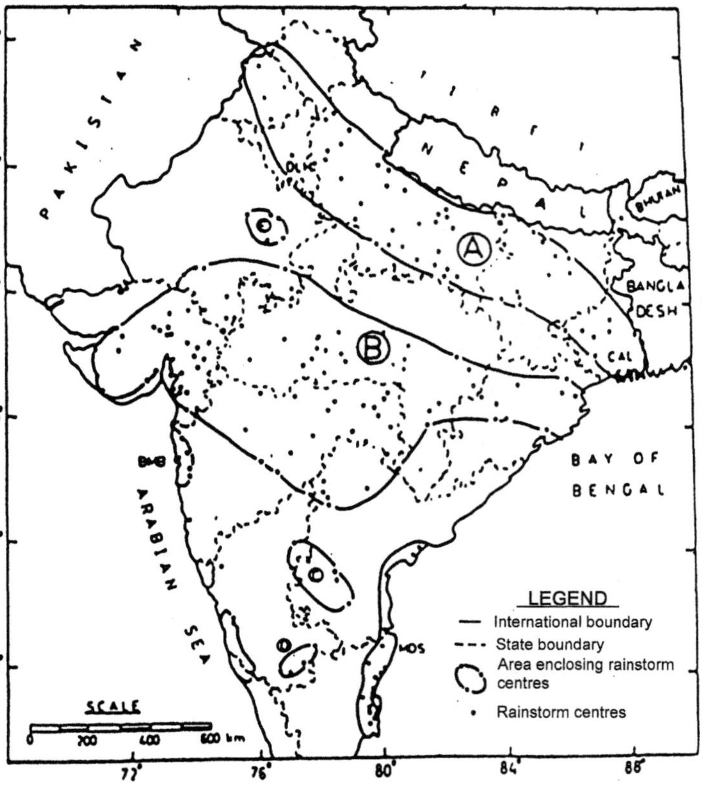

Figure 5. Zones of severe rainstorms over India with their rainstorm centers.

(Figure 5). Slow moving and stationary cyclonic disturbances cause exceptionally heavy falls of rain over the regions they pass through. As a result of heavy rainfall severe floods occur in the rivers of the region. These floods generally are termed as single event flood. The floods in northern and central Indian rivers occur mostly due to heavy rainfall caused by the slow movement of these disturbances. The deltaic areas of Orissa, Andhra Pradesh and also Tamil Nadu (during winter monsoon) suffer frequently from the floods caused by these Bay cyclonic disturbances at the time of landfall.

In case of northeastern rivers, viz. the Brahmaputra and its tributaries, the Barak, etc., frequent major floods occur during the monsoon period of June to September. These floods are caused by heavy rainfall associated with 'Break' monsoon situations or recurving monsoon depressions from the Bay of Bengal (Dhar and Nandargi, 2000).

5.2. 'BREAK' MONSOON SITUATION

This weather situation is a speciality of the Indian summer monsoon. This happens when the axis of the seasonal monsoon trough, which normally passes through Sri Ganganagar (29°55′N/73°53′E), Delhi (28°35′N/77°12′E), Kanpur (26°26′N/80°22′E), Patna (25°37′N/85°10′E) and then to Calcutta (22°32′N/88°20′E), shifts northwards from its normal position and lies close to the foothills of the Himalayas both on surface and 850 hPa synoptic weather charts. This particular situation results in causing heavy rainfall over the north-eastern and central Himalayas and their adjoining plain areas (Dhar *et al.*, 1984b), while the rest of the country is reeling under drought conditions with more or less no rain.

During 'Break' situations, due to heavy rainfall over Himalayan and sub-Himalayan regions, floods occur in the rivers of northeast and central Himalayas, viz., the Brahmaputra and its tributaries, the Teesta, the Kosi, the Gandak, the Ghagra, the Kamla Balan, the Bagmati, and the Rapti. Due to steep slopes of the Himalayas, the run-off is heavy in these rivers and flood waters bring along with them heavy silt from the fragile hill slopes of the Himalayas which not only augment the flood waters in the river channels but also cause bank erosion in the plains-reach of these rivers. This is a paradoxical situation so far as the Indo-Gangetic plains are concerned as floods occur in most of the Himalayan rivers while there is practically no rain in the plains reach of these rivers. It has however, been seen that 'Break' situations last very rarely for more than a week at a time and they generally occur in July and August months of summer monsoon season.

5.3. EL-NINO AND LA-NINA PHENOMENA

Association of El-Nino and La-Nina episodes have been found to be related with monsoon rainfall over India and in a way are responsible for causing either droughts (El-Nino +ve) conditions or good-monsoon i.e., floods (La Nina) over India. But it has been seen that all El-Nino episodes do not cause droughts and in a similar way all La-Nina episodes do not cause excess rainfall (Kulshrestha, 1997). The effect of these events is still debatable and considerable research work is going on at present to find their linkages with the Indian summer monsoon rainfall (Das, 1999).

According to Kripalani and Kulkarni (1997) there are short term fluctuations of seasonal rainfall and due to that there are epochs of above and below normal rainfall over the country during the monsoon period. If on these, epochal behaviour is superimposed by external forcing of El-Nino/La-Nina and the two are in phase, major extreme events of rainfall occur during the summer monsoon months. In other words, due to phase-locking between epochal variability and the external forcing due to El-Nino and La-Nina episodes during a summer monsoon season, severe drought conditions or severe wet monsoon (or floods) occur in the country.

It is also seen that besides these major synoptic situations, floods in the Indian region are caused by the incidence of heavy to very heavy rainfall, which are associated with any one or combination of more than one of the following synoptic systems:

(a) Passage of low pressure systems (LPS) or monsoon lows through the country,

(b) Active monsoon conditions prevailing over a region for a fairly good number of days,
(c) Off-shore vortices along the west coast of India,
(d) Mid-latitude westerly systems moving from west to east, and
(e) Mid-tropospheric cyclonic circulations over western region of the country.

The strength of the summer monsoon in any one year depends upon the frequency of the above meteorological systems moving through the country and the tracks they follow. Apart from these systems, orography of the country (Figure 1) also plays an important role in causing heavy rainfall during the monsoon months. As an example, the heavy rainfall over the Western Ghats and along the Khasia Jantia hills of northeast India is mainly due to the orography of these regions in making the prevailing highly moist monsoon winds to rise up to great vertical heights and making them to shed their moisture over the steep hilly slopes causing, thereby, quick runoff and consequent floods. Himalayan slopes of Assam in north Assam and Bhutan too receive very heavy rainfall during the summer monsoon months.

5.4. MAN-MADE FACTORS CAUSING FLOODS

Besides the above mentioned meteorological situations, there are some man-made factors which are also responsible for causing serious floods in different parts of the country.

(a) Indiscriminate destruction of forests and vegetal cover in the upper reaches of river basins over a long period of time. This results in increasing surface-runoff and decreasing absorptive capacity of the soil in the catchment areas of rivers. Due to increased runoff from mountain slopes, widespread gully and sheet erosion takes place, which decrease the capacity of the rivers to carry increased runoff because of siltation of river beds.
(b) Improper agricultural practices, such as shifting cultivation also help in causing soil erosion and increased surface runoff.
(c) Obstructions to free flow of water in rivers, due to embankments, over flow of dams, river bunds and narrow bridge-openings also help in causing floods.

6. Types of Floods

As stated above, floods can be caused by various factors besides weather systems which cause heavy rainfall. Generally the type of flood depends upon that very factor which is responsible for causing it. The various types of floods are briefly mentioned below.

6.1. SINGLE EVENT FLOODS

This is the most common type of flooding in which widespread heavy rains of a longer duration of 2 to 3 days over a drainage basin results in severe floods. Such heavy rains are associated with cyclonic disturbances, such as cyclonic storms, low pressure areas, depressions, etc., during the summer monsoon season when the moisture content of air is very high. In the event that these disturbances move slowly, they cause concentrated heavy rainfall, say 10 to 15 mm/h over an area resulting in severe floods over an area. In India most of the floods are of this type.

6.2. FLASH FLOODS

These are frequently associated with violent, convectional storms of short duration and cause heavy rainfall over an area in a short duration. Such floods occur mostly in hilly areas and cause local streams and rivulets to rise suddenly destroying bridges, buildings, roads and crops. The Vamsadhara flash flood of September, 1980 is a well known example of this type. It was studied in detail by Dhar *et al.* (1981b) and it caused huge loss of life and property in Orissa and Andhra Pradesh due to on-rush of flood wave which moved with considerable speed within a short duration.

6.3. FLOODS CAUSED BY SUCCESSIVE WEATHER EVENTS

There are occasions when floods occur in river basins by the movement of successive weather disturbances, like low pressure areas or depressions, one after another with hardly a gap of one or two days. In such situations there is a continuous spell of rain (not necessarily very heavy) which causes gradual rise of water in the rivers of the region. Such floods are of long duration and often occur in Central Indian rivers of Narmada, Tapi, Mahanadi, etc. as these basins fall within the broad width of storm tracks of weather systems that originate from the Bay of Bengal and move over central India in a westerly to northwesterly direction.

Nandargi and Dhar (1998) showed that two rainstorms of July 1927, i.e., 23–29 July and 27–30 July occurring with hardly one-day gap between the two caused heavy rainfall and floods in the two distant parts of India, i.e., river Mahi and its tributaries in Gujarat and river Baitarni and its tributaries in Orissa states. Similarly, during 30 August to 6 September 1970, two upper air cyclonic circulations and a deep depression, created active monsoon conditions over the Narmada basin resulting in highest ever recorded flood at Garudeshwar site on 6 September, 1970 (Table IV).

6.4. SEASON FLOODS

These are floods which occur during different seasons. As stated earlier, summer monsoon season experiences large number of floods, since major storm activity occurs during this season. The southern half of the Indian peninsula experiences floods mostly during the winter monsoon season.

The above mentioned floods are caused directly by heavy rainfall over a drainage basin. However, floods can also occur due to unusually high water levels of lakes in which rivers fall. This often happens in the case of Jhelum River when the Walur Lake water level is higher than the level of water in the Jhelum (Dhar and Nandargi, 1992a).

6.5. OTHER CAUSES OF FLOODS

Sometimes flood events may be caused by events other than rainfall as is clear from the following examples:

(a) Coastal floods: These floods are generally caused by storm surges (or tidal waves) which occur in the coastal belt at the time of storm crossing the coast. The surge is the outcome of piling of sea water against the coast due to strong winds generated by a cyclonic storm. The surface waters are driven towards the coast where due to shallow water the return flow is retarded by the frictional force of the sea bed. If the surge takes place near the mouth of a river falling into the sea, the river flow is held up due to the surge which result in severe flooding over and near the coastal areas. Such coastal floods often occur along the east coast of India especially where east flowing rivers fall into the Bay of Bengal during summer and winter monsoon months due to the activity of tropical cyclones/depressions in the Bay of Bengal.

According to Rama Sastry (1984), storm surges are responsible for nearly 90% of lives lost in a tropical cyclone. During the November 1977 Andhra cyclone, the peak storm surge, when the cyclone crossed the coast near Chirala on 19 November evening, was estimated to be 5.5 meters high. This enormous mass of water washed away many coastal villages killing nearly 15,000 human lives in and near coastal areas of Krishna district in Andhra Pradesh.

The other example of this type of coastal flooding is the 8–13 November 1970 cyclonic storm which struck coastal Bangladesh. It created 3 to 10 m high storm surge which resulted in the death of over 300,000 lives and damage to property worth billions of rupees (Sikka, 2000). Similarly, in India, the recent Orissa super cyclone of 28–31 October, 1999 caused tremendous damage to life and property along the Orissa coast. This cyclone produced very heavy rainfall and flooding along the coastal belt, Paradeep station recorded a rainfall of 55 cm in 24 hours.

(b) Floods caused by dam failures: Failure of dams, or collapse of obstructions to the flow of water in a river by glacial tongues moving down the mountain slopes or occurrence of land slides across rivers flowing through hills, cause serious floods in downstream areas when these obstructions give way due to pressure of accumulated water upstream of the obstructions. In 1979, the collapse of Machhu dam in Saurashtra caused huge loss of about 10,000 human lives in Morvi town alone besides tremendous destruction of live stock, property and agricultural lands (Dhar *et al.*, 1981c). Collapse of Khadkwasla-Panshet dams in July, 1961, in Pune is another example of dams giving way due to heavy rush of flood waters upstream of the dam. One of the largest floods which occurred in the Indus River originated from landslide dam failure in 1841 (Kale, 1996).

Sometime artificial lakes are formed due to land slides, blocking the flow of a river or a glacier tongues invading the river bed by sliding from steep mountain slope causing artificial lakes in the river bed which on bursting causes devastation and destruction downstream of the dam. The 1926 and 1929 floods in the Shyok river in the Ladakh region of the country were caused by bursting of artificial lakes created by invading glacier tongues into the river bed (Kale, 1998).

(c) Eustarine floods: Estuaries are the only portions of coastline where the normal tidal currents meet a concentrated seaward flow of fresh water in a river. The interaction between the seaward flow of fresh water in a river and landward flow of saline water from sea, during high tides may result in opposing land water flow causing a ‘wall’ of water. Sometime the funnel shape characteristic of many estuaries causes increase in high water level in the upper narrow reaches of rivers. These types of floods are mostly experienced in deltaic areas of rivers along the coasts and are not considered as serious floods.

(d) Floods due to snow or glacier ice melting: Occasionally Himalayan rivers experience minor to medium floods due to sudden rise of air temperature which causes quick melting of snow and glacier ice in the upper reaches of rivers. Floods also occur due to heavy rains over snow fields having deep snow cover. Such floods occur during the months of April to June in the plain areas adjoining Himalayas. Floods of this type are rather uncommon and occur mostly in Himalayan valleys and do not cause large-scale destruction or damage in the plains. In addition, cloud bursts have often been found to cause severe floods over small catchments of streams but these floods hardly last only for a short duration, say a couple of hours.

But all these floods are of medium or minor character and cannot cause serious floodings. In the present study, floods that occurred in Indian rivers (major and medium) during the summer monsoon seasons of the 15-year period (1986–2000) are discussed. This study will be useful to planners, environmentalist and flood forecasters for undertaking remedial measures to fight recurring floods in flood-prone regions of the country.

7. Floods in Indian Rivers during the Summer Monsoon Seasons of the Period 1986–2000

Floods of various magnitudes in one or the other parts of the country are an annual feature of this country. As stated earlier, during the summer monsoon season, passage of tropical disturbances like cyclonic storms, depressions, low pressure areas, etc. are responsible for causing heavy rainfall which result in causing floods in the different rivers of the country. Dhar and co-workers (1959, 1966, 1975, 1981d, 1986, 1992b, 1993, 1994, 1998) have been studying these floods regularly with the sole aim of understanding their yearly space-time distribution over the country. In the present study, the recent 15-year flood data have been analyzed and the results are presented.

Relevant flood data for the present study were collected mainly from the following publications:

(a) Weekly Flood News Letters (WFNLs) issued by the Central Water Commission (1986–2000), New Delhi during every flood season, which broadly coincides with the summer monsoon season of India, and

(b) Rainfall and other meteorological data published by the India Meteorological Department (1986–2000) and the flood data supplied by the State Governments.

The gauge/discharge (G/D) sites on these rivers have been installed by the concerned State Governments through which these rivers flow, in consultation with Central Water Commission (CWC), New Delhi, which is the nodal agency responsible for the development of water resources of this country.

Table I gives the frequency of floods in each year of the 15-year period from 1986 to 2000. Examination of Table I shows that maximum number of rivers (i.e., about 40) were in floods in 1996 and 1998 and the average monsoon rainfall over the country was +2% and +6% above the normal respectively during these two monsoon seasons. However, in the best monsoon season of 1988, when monsoon rainfall was +18% above the normal, only 32 rivers were in flood. In the worst-drought year of 1987, when rainfall was −18% below normal, floods occurred in 30 rivers. It is also seen that during the 15-year period the number of rivers which experienced floods, varied from 25 (in 1992) to 40 (in 1996 and 1998), although the variation in average summer monsoon rainfall over the entire country was from −18% to +18%.

It is therefore seen from Table I, that the number of rivers affected by floods does not depend upon the magnitude of monsoon rainfall received by the country as a whole. A monsoon season having large negative percentage departure of rainfall from the normal may record the same number of floods as the one having large positive departure of rainfall, as happened in the years 1987 and 1988.

It is also seen that highest number of floods (i.e., 429) were recorded during the 1998 flood season of which 136 were major floods, i.e., floods stood more than 1 m above their D.L.

8. Catalogue of Highest Recorded Floods in the Indian Rivers

A comprehensive catalogue of highest recorded floods of all Indian rivers (numbering about 69) and their G/D sites (numbering about 235) has been prepared by

Table I. Frequency of floods in different rivers and years (1986–2000)

	1986	1987	1988	1989	1990	1991	1992	1993	1994	1995	1996	1997	1998	1999	2000
Moonsoon rainfall over contiguous Indian areas as % of normal	−16%	−18%	+18%	−1%	+7%	−7%	−8%	+1%	+11%	0	+2%	+2%	+6%	−4%	−7%
Total No. of rivers and their G/D sites (in brackets) affected by floods	28 (57)	30 (58)	32 (68)	28 (46)	37 (66)	34 (62)	25 (51)	34 (61)	33 (71)	33 (57)	40 (75)	38 (56)	40 (70)	34 (63)	35 (69)
Total No. of floods recorded at G/D sites on different rivers	192	311	329	181	291	260	115	290	244	279	277	193	429	334	320
Total No. of major floods (i.e., flood level 1 m & above D.L.) recorded at G/D sites	55 (24)	108 (35)	96 (35)	29 (14)	51 (23)	62 (36)	16 (12)	48 (24)	69 (27)	55 (22)	77 (34)	39 (20)	136 (37)	88 (28)	63 (25)

Figures in brackets indicate the number of gauge/discharge (G/D) sites affected by floods.

Nandargi (1996) using all available flood data up to 1994. However, in the present study, on the basis of recent 15-year flood data from 1986 to 2000, a consolidated list of highest recorded floods in major rivers of the country at different G/D sites has been prepared updating the previous flood records. This list is given in Table II.

From a perusal of Table II, it is seen that 53 major rivers of India recorded the highest floods at 118 G/D sites. The distribution of these major rivers and their G/D sites in different flood prone regions of the country show that 28 sites on 19 rivers in NE India, 66 sites on 21 rivers in north India and 24 sites on 13 rivers in Peninsular India, recorded floods above their respective danger levels (DLs) up to a height of 18.10 m at different G/D sites. Figure 6 shows the frequency distribution of floods at different G/D sites on different river systems of the country during the 1986 to 2000 flood seasons.

9. Flood-Prone River Systems/Regions of the Country (1986–2000)

In order to know which state/region of the country is prone to floods and in which months of the monsoon period (from June to September) their frequency is the highest, the monthly distribution of these floods in different states/regions during the period 1986–2000 has been prepared and is given in Table III. From a perusal of this table, it can be seen that monsoon months of July, August and September are the worst flood affected months of the country. Comparatively very few floods have occurred in the pre- and post-monsoon months of May and October.

It is also seen that the states, worst affected by floods are Assam (in the Brahmaputra basin), and Bihar (in the Ganga basin). To a lesser extent, states of Uttar Pradesh and West Bengal experience floods in some of their major rivers. It has also been observed that very few floods have occurred in the rivers of Peninsular India.

It is seen from Figure 6 that frequency of floods at G/D sites is the highest in northeast Indian rivers and next to it are the G/D sites of north Indian rivers. In the following sections a brief description of floods experienced in different river systems is given.

9.1. FLOODS IN BRAHMAPUTRA RIVER SYSTEM (NORTH-EAST INDIA)

The Brahmaputra basin is one of the largest river basin in the northeast region of India i.e., Assam and neighbouring states. This is the region of highest rainfall in India because of its geographical location and orography, which contribute the maximum flow in this river. The Brahmaputra river originates in the semi-arid region of south Tibet, but due to its passage through the rainiest region of India it experiences year after year severe floods, which bring death and devastation not only to Assam but also to the neighbouring country of Bangladesh. The main meteorological situations which cause heavy rainstorms over the Brahmaputra basin are due to the shifting of the eastern end of the seasonal monsoon trough to the

Table II. List of highest recorded floods in the major Indian rivers at different Gauge/Discharge sites (1986-2000)

Sr. No.	River	Gauge site	State	Danger level (m)	Highest flood level (m)	Date of occurrence	Difference in flood level (m)	Total No. of floods (major floods, i.e., ≥1 m) (1986–2000)
				(A) Northeast India				
(a) Main Brahmaputra River:								
1.	Brahmaputra	Dibrugarh	Assam	104.24	106.41	04.09.98	2.17	172 (54)
		Nematighat	Assam	85.04	87.32	11.07.91	2.28	130 (21)
		Tezput	Assam	65.23	66.58	27.08.88	1.35	75 (5)
		Guwahati	Assam	49.68	52.23	08.08.2000	2.55	46 (5)
		Goalpara	Assam	36.27	37.29	29.08.88	1.02	15 (1)
		Dhubri	Assam	28.50	30.18	28.08.88	1.68	111 (13)
(b) Northern tributaries of the Brahmaputra River:								
(from east to west)								
2.	Subansiri	Badatighat	Assam	82.53	83.84	05.09.98	1.31	39 (3)
3.	Jaibareilly	N.T. Rd. Crossing	Assam	77.00	78.04	11.07.98	1.04	98 (1)
4.	Puthimari	N.H. Crossing	Assam	51.81	56.46	20.06.93	4.65	70 (23)
5.	Beki	Road Bridge	Assam	44.38	46.16	04.08.2000	1.78	191 (22)
6.	Pagladiya	N.T. Rd. Crossing	Assam	52.75	54.40	02.08.2000	1.65	29 (10)
7.	Manas	N.H. Crossing	Assam	47.56	49.76	11–17.07.96	2.20	107 (38)
8.	Sankosh	Golakganj	Assam	29.51	30.58	22.07.93	1.07	84 (2)
9.	Burhi Dihing	Naharkatia	Assam	120.40	121.11	13.08.95	0.71	5(–)
		Khowang	Assam	102.11	103.85	26.08.88	1.74	92 (26)
10.	Desang	Nanglamoraght	Assam	94.46	96.45	07.09.98	1.99	48 (17)
11.	Dikhaw	Sibsagar	Assam	93.30	94.66	24.06.90	1.36	13 (3)

Table II. Continued

Sr. No.	River	Gauge site	State	Danger level (m)	Highest flood level (m)	Date of occurrence	Difference in flood level (m)	Total No. of floods (major floods, i.e., ≥1 m) (1986–2000)
12.	Dhansiri	Golaghat	Assam	89.50	91.30	11.10.86	1.80	21 (3)
		Numaligarh	Assam	77.42	79.68	11.10.86	2.26	150 (47)
13.	Kopili	Kampur	Assam	60.50	61.56	11.10.86	1.06	31 (4)
		Dharamtul	Assam	56.00	57.54	09.07.88	1.54	34 (8)
(c) Rivers joining the Brahmaputra in Bangla Desh:								
14.	Barak	Annapurnaghat (Silchar)	Assam	19.83	21.77	11.10.86	1.94	32 (7)
15.	Katakhal	Matizuri	Assam	20.27	22.48	12.07.97	2.21	56 (31)
16.	Raidak	Tufanganj	West Bengal	35.30	36.16	21.07.93	0.86	15 (–)
17.	Torsa	Ghugumari (Cooch Behar)	West Bengal	40.41	41.10	11.08.87	0.69	5 (–)
18.	Jaldhaka	Mathabanga	West Bengal	48.70	49.42	10.08.87	0.72	7 (–)
19.	Teesta	Dmohini Rd. Bridge	West Bengal	85.86	86.15	12 & 13.07.96	0.29	5 (–)
		Baltali		65.95	66.00	13.07.96	0.05	1 (–)
				(B) North India				
(a) Main Ganga River: (from west to east)								
20.	Ganga	Hardwar	Uttar Pradesh	294.00	295.35	03.08.97	1.35	11 (3)
		Narora (Downstream)	Uttar Pradesh	178.00	178.86	19.07.86	0.46	2 (–)
		Ankinghat	Uttar Pradesh	124.00	124.11	29.08.98	0.11	2 (–)
		Dalmau	Uttar Pradesh	99.36	99.58	31.08.98	0.22	4 (–)
		Allahabad (Ghatnag)	Uttar Pradesh	84.73	85.73	26–27.08.96	1.0	4 (1)

Table II. Continued

Sr. No.	River	Gauge site	State	Danger level (m)	Highest flood level (m)	Date of occurrence	Difference in flood level (m)	Total No. of floods (major floods, i.e., ≥1 m) (1986–2000)
20.	Ganga	Phaphamau	Uttar Pradesh	84.73	86.35	25–27.08.96	1.62	6 (3)
	(continued)	Mirzapur	Uttar Pradesh	77.72	78.69	26–27.08.96	0.97	5 (–)
		Varanasi	Uttar Pradesh	71.26	72.54	29–31.08.96	1.28	9 (2)
		Gazhipur	Uttar Pradesh	63.11	65.10	29.08–01.09.96	1.99	21 (7)
		Ballia	Uttar Pradesh	57.62	64.26	5–11.09.96	6.64	68 (31)
		Buxar	Bihar	60.32	61.03	29.08–01.09.96	0.71	12 (–)
		Manger	Bihar	52.00	52.67	10.09.95	0.67	2 (–)
		Dighaghat (Patna)	Bihar	50.45	51.71	02.09.91	1.26	16 (2)
		Gandhighat (Patna)	Bihar	48.60	50.12	13.09.87	1.52	59 (8)
		Hatidah	Bihar	41.76	42.67	14 & 15.08.94	0.91	45 (–)
		Monghyr	Bihar	39.33	39.58	15.09.87	0.25	3 (–)
		Bhagalpur	Bihar	33.68	33.97	06.09.98	0.29	9 (–)
		Colgong	Bihar	31.09	32.83	06.09.98	1.74	68 (14)
		Farakka	West Bengal	22.25	25.12	07.09.98	2.87	118 (62)
(b) Northern tributaries of the Ganga River:								
(from west to east)								
21.	Sai	Rai Bareilly	Uttar Pradash	101.00	102.07	20 & 21.08.88	1.07	3 (1)
22.	Ghagra	Elgin Bridge	Uttar Pradesh	106.07	106.77	19.08.88	0.70	64 (–)
		Ayodhya	Uttar Pradesh	92.73	93.64	21.08.88	0.91	70 (–)
		Turtipar	Uttar Pradesh	64.01	65.95	29.08.98	1.94	92 (7)
		Darauli	Uttar Pradesh	60.82	61.82	18.08.98	1.00	17 (1)
		Gangpur Siswan	Uttar Pradesh	57.04	57.97	20.09.93	0.93	26 (–)
		Chhapra	Uttar Pradesh	53.68	54.75	11.08.88	1.07	6 (1)

Table II. Continued

Sr. No.	River	Gauge site	State	Danger level (m)	Highest flood level (m)	Date of occurrence	Difference in flood level (m)	Total No. of floods (major floods, i.e., ≥1 m) (1986–2000)
23.	Rapti	Balrampur	Uttar Pradesh	104.62	105.23	11.09.2000	0.61	8 (–)
		Bansi	Uttar Pradesh	84.90	85.77	14.09.2000	0.87	18 (–)
		Birdghat (Gorakhapur)	Uttar Pradesh	74.98	77.38	24.08.98	2.40	38 (15)
24.	Gandak	Khadda	Uttar Pradesh	96.00	96.90	02.08.2000	0.90	47 (–)
		Rewaghat	Uttar Pradesh	54.41	55.34	18.09.86	0.93	6 (–)
25.	Burhi Gandak	Lalbegiaghat	Bihar	63.20	64.74	14.08.87	1.54	15 (6)
		Sikandarpur	Bihar	52.53	54.29	15.08.87	1.76	23 (4)
		Samastipur	Bihar	46.02	49.38	16.08.87	3.36	46 (23)
		Rossera	Bihar	42.63	46.35	16.08.87	3.72	55 (26)
		Khagaria	Bihar	36.58	38.45	18.09.87	1.87	61 (20)
26.	Bagmati	Benibad	Bihar	48.68	49.72	06.07.99	1.04	104 (2)
		Hayaghat	Bihar	45.72	48.96	14.08.87	3.24	54 (20)
27.	Adhwara Group	Kamtur	Bihar	50.00	51.38	18–24.07.96	1.38	67 (7)
		Ekmighat	Bihar	46.94	49.27	14.08.87	2.33	57 (14)
28.	Kamala Balan	Jhanjharpur	Bihar	50.00	52.67	12.08.87	2.67	108 (35)
29.	Kosi	Basua	Bihar	47.75	48.94	28.07.98	1.19	75 (1)
		Baltara	Bihar	33.85	36.66	20.07.98	2.81	164 (76)
		Kursella	Bihar	30.00	32.04	06.09.98	2.04	89 (31)
30.	Mahananda	Dhengraghat	Bihar	35.65	36.85	11.09.91	1.20	61 (6)
		Jawa	Bihar	31.40	35.85	15–21.08.96	4.45	57 (10)

Table II. Continued

Sr. No.	River	Gauge site	State	Danger level (m)	Highest flood level (m)	Date of occurrence	Difference in flood level (m)	Total No. of floods (major floods, i.e., ≥1 m) (1986–2000)
(c) Southern tributaries of the Ganga River:								
(c1) *Main Yamuna River*:								
31.	Yamuna	Mawi	Delhi	230.85	231.94	07.09.95	1.09	6 (1)
		Delhi Rly. Bridge	Delhi	204.83	206.92	27.09.88	2.09	32 (3)
		Mathura	Uttar Pradesh	165.20	167.38	12.09.95	2.18	31 (5)
		Etawah	Uttar Pradesh	121.00	122.89	25–27.08.96	0.97	3 (–)
		Auraiya	Uttar Pradesh	110.00*	116.63	30.07.86	6.63	7 (6)
		Kalpi	Uttar Pradesh	108.00	112.75	23.08.96	4.75	7 (6)
		Hamipur	Uttar Pradesh	103.63	107.11	23.08.96	3.48	10 (5)
		Chollaghat	Uttar Pradesh	100.00	102.79	24–28.08.96	2.79	12 (7)
		Naini	Uttar Pradesh	84.73	86.14	25–28.08.96	1.41	5 (3)
(c2) *Tributaries of the Yamuna River*:								
32.	Betwa	Mohana	Uttar Pradesh	122.66	124.02	19.09.93	1.36	7 (2)
		Shahjina	Uttar Pradesh	104.55	106.52	22–28.08.96	1.97	5 (1)
33.	Ken	Banda	Uttar Pradesh	104.00	113.24	14.09.92	9.24	24 (18)
(c3) *Other rivers*:								
34.	Punpun	Sripalpur	Bihar	50.60	53.50	15.09.87	2.90	69 (52)
35.	Sone	Koelwar	Bihar	55.52	58.37	12.09.87	2.85	20 (13)
		Maner	Bihar	52.00	53.60	15.09.87	1.60	35 (8)
36.	Ajoy	Cheropara	West Bengal	39.42	39.69	28.08.87	0.27	1 (–)
37.	Mayurkoshi	Narayanpur	West Bengal	27.99	28.28	20.10.99	0.29	2 (–)
38.	Mundeshwari	Harinkhola	West Bengal	12.80	14.45	24.09.2000	1.65	14 (4)

Table II. Continued

Sr. No.	River	Gauge site	State	Danger level (m)	Highest flood level (m)	Date of occurrence	Difference in flood level (m)	Total No. of floods (major floods, i.e., ≥1 m) (1986–2000)
39.	Kangsebati	Mohanpur	West Bengal	25.73	26.72	07.09.95	0.99	7 (–)
(d) Other rivers of North India:								
40.	Mahi	Wanakbori	Gujarat	71.93	74.68	24.08.90	2.75	7 (2)
				(C) Central India				
(a) East-central rivers:								
41.	Baitarni	Anandpur	Orissa	38.36	39.50	07.08.96	1.14	6 (1)
		Akhupada	Orissa	19.20	20.39	05.08.89	1.19	4 (1)
42.	Brahmini	Jenapur Exp. Way	Orissa	23.00	23.32	14.08.91	0.32	1 (–)
43.	Burhabalang	N.H.5 Bridge	Orissa	8.13	8.30	22.08.97	0.17	2 (–)
44.	Mahanadi	Alipingal Devi	Orissa	11.76	12.51	15.08.91	0.75	4 (–)
		Nimpara	Orissa	10.76	11.19	15.08.91	0.43	7 (–)
45.	Rushikulya	Purushottampur	Orissa	16.83	17.65	22.08.97	0.82	3 (–)
46.	Subarnarekha	Rajghat	Orissa	10.36	12.19	07.08.97	1.83	15 (6)
47	Vamsadhara	Gunupur	Orissa	84.00	86.30	29.07.91	2.30	2 (1)
		Kashinagar	Orissa	54.60	55.64	29.07.91	1.04	4 (1)
(b) Narmada River:								
48.	Narmada	Mandla	Madhya Pradesh	437.80	438.68	24.08.91	0.88	3 (–)
		Hoshangabad	Madhya Pradesh	293.83	298.25	21.09.99	4.42	9 (5)
		Garudeshwar	Madhya Pradesh	31.09**	38.10	07.09.94	7.01	7 (3)
		Broach	Madhya Pradesh	7.31	12.00	08.09.94	4.69	27 (19)
(c) Tapi River:								
49.	Tapi	Surat	Gujarat	9.50	11.50	17.09.98	1.60	1 (1)

Table II. Continued

Sr. No.	River	Gauge site	State	Danger level (m)	Highest flood level (m)	Date of occurrence	Difference in flood level (m)	Total No. of floods (major floods, i.e., ≥1 m) (1986–2000)
(d) Main Godavari River:								
50.	Godavari	Dummagudam	Andhra Pradesh	55.00	58.95	24.08.90	3.95	7 (2)
		Bhadrachalam	Andhra Pradesh	48.77	53.61	24.08.90	4.84	8 (7)
		Kunavaram	Andhra Pradesh	40.32	48.10	25.08.90	7.78	10 (9)
		Rajamundry	Andhra Pradesh	19.51	19.83	25.08.90	0.32	1 (–)
		Dowleswaram	Andhra Pradesh	16.08	17.40	26.08.90	1.32	9 (1)
(d1) *Tributaries of the Godavari River*:								
51.	Wainganga	Bhandara	Maharashtra	244.50	248.84	07.09.94	4.34	10 (5)
		Paoni	Maharashtra	232.50	236.11	08.09.97	3.61	10 (4)
52.	Indravati	Jagdalpur	Madhya Pradesh	540.80	543.80	29.07.92	3.00	13 (7)
(e) South India:								
53.	Tungabhadra	Mantralayam	Andhra Pradesh	312.00	312.93	19.07.94	0.93	1 (–)

*Danger level of Auraiya G/D site on the Yamuna River has been changed since 1991, to 113.00 m.

**Danger level of G/D site, Garudeshwar on the Narmada River has been changed since 1997, to 31.09 m.

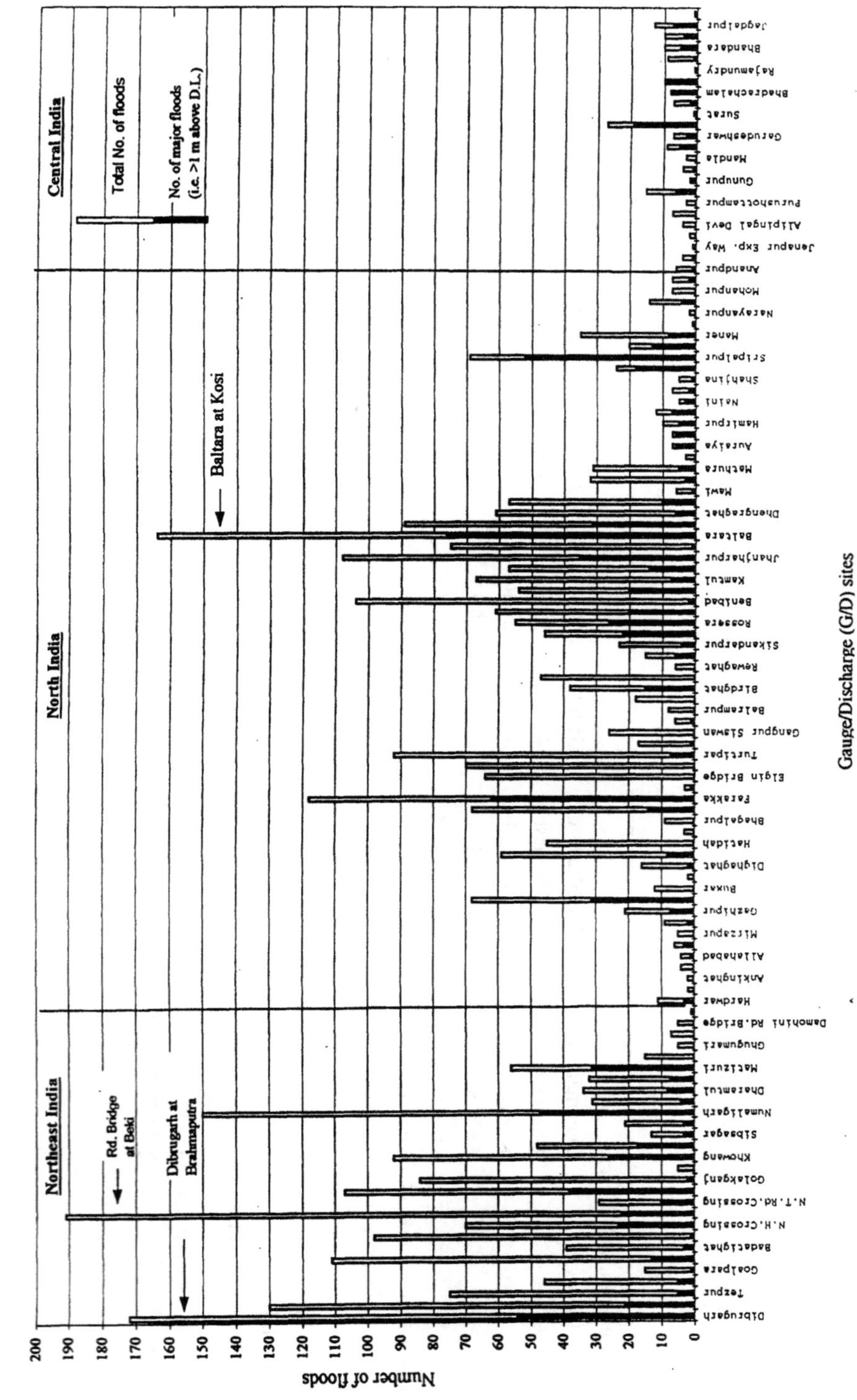

Figure 6. Number of severe floods recorded at different G/D sites on the major rivers of India (1986–2000).

Table III. Statewise and monthwise distribution of floods at different G/D sites (1986–2000)

Sr. No.	Month/ State	May	June	July	August	September	October	Total
1.	Andhra Pradesh			8(3)	16 (9)	12 (7)		36 (19)
2.	Assam and neighbourhood	22 (7)	275 (41)	572 (126)	449 (107)	305 (49)	44 (15)	1649 (345)
3.	Bihar		12	271 (54)	611 (173)	523 (152)	15	1432 (379)
4.	Delhi			7	20 (1)	11 (3)		38 (4)
5.	Gujarat			10 (4)	19 (12)	13 (9)		42 (25)
6.	Madhya Pradesh			9 (6)	7 (2)	8 (3)	1 (1)	25 (12)
7.	Maharashtra			10 (5)	4	6 (4)		20 (9)
8.	Orissa		2	5 (3)	31 (6)	7 (1)	3	48 (10)
9.	Uttar Pradesh		1	96 (24)	297 (58)	184 (43)	2	580 (125)
10.	West Bengal		3 (1)	30 (3)	67 (28)	69 (34)	6	175 (66)
	Total	22 (7)	275 (42)	1018 (228)	1521 (396)	1138 (305)	71 (16)	4045 (994)

Figures in brackets indicate major floods (i.e., 1 m and more above the D.L.).

foot-hills of the Himalayas in the north, 'Break' monsoon situations and active monsoon conditions prevailing over the region.

It has been seen that the highest flood deviations from D.L. at G/D sites over the main Brahmaputra river varies from 1 m to 2.55 m during 1986-2000. The two G/D sites at Dibrugarh and Nematighat in upper Assam have been experiencing high floods year after year as well as maximum frequency of floods of the order of 130 to 172 during the 15-year period (see Table II).

Among the Brahmaputra's 12 major tributaries in Assam, the Puthimari river has recorded the highest deviation from D.L. of 4.65 m in June 1993 while the Beki river has experienced the highest frequency of floods of the order of 191 in 15-year period.

9.2. FLOODS IN GANGA SYSTEM AND ITS HIMALAYAN TRIBUTARIES (NORTH INDIA)

This is the largest river system of the north Indian region. The river originates from the Gangotri glacier which is located at an elevation of 7010 m a.s.l. in the Garhwal Himalayas. The Ganga river from its source to its outfall into the Bay of Bengal flows through 8 north Indian states covering an area of about 861,404 km^2 with its total length of about 2,525 km.

(a) *Ganga – The Main River*

Frequency of floods in the Ganga at its G/D sites generally increases after the river enters Bihar from Uttar Pradesh. In other words, from Ballia onwards till it reaches Farakka, there is a general tendency for floods to increase in frequency in this section of the river which is mainly due to drainage congestion. This is evident by the flood frequencies at Ballia, Gandhighat (Patna), Colgong and Farakka G/D sites (see Table II). The reason for this is that because in this section of the river maximum number of tributaries of the Ganga from the north as well as from the south join the main Ganga river carrying with them huge volumes of flood discharge and silt from their respective catchments during the period of monsoon months of July, August and September.

(b) *Ganga – Its Northern Tributaries*

Among the northern tributaries of the Ganga having large catchment areas in the Himalayas, (viz., the Ghagra, the Gandak and the Kosi), the Kosi and the Ghagra appear to be more vulnerable to frequent floods. Among the rivers rising from the foot-hills of the Himalayas, the Kamla Balan has been experiencing frequent floods almost every year.

(c) *Ganga – Its Southern Tributaries*

Tributaries of the Ganga from the south, except the Punpun, have not been that effective in causing floods as the tributaries from the north. Rivers, Sone and Ken have been experiencing less number of floods compared to the Punpun.

The westernmost tributary of the Ganga, viz., the Yamuna, has been experiencing occasional floods in the section from Delhi to Mathura and also beyond this section, floods have been occurring in this river less frequently till it joins the Ganga at Allahabad in Uttar Pradesh (see Figure 6).

Comparative study of frequency of floods at the terminal sites of the two major rivers, i.e., the Brahmaputra at Dhubri and the Ganga at Farakka have shown that their flood frequencies are more or less of the same order (i.e., 111 and 118 respectively) (see Table II) even though rainfall magnitudes received by each of these two basins during monsoon period are quite different. Perhaps, the pattern of rainfall distribution, long length of these two rivers and their tributaries as well as the rugged orography of their upper reaches in the Himalayas and trans-Himalayas and cross section of these rivers at their terminal G/D sites are responsible for this unique feature of their flood frequencies at their respective terminal sites of Dhubri and Farakka.

9.3. FLOODS IN CENTRAL INDIA

There are about 13 major and medium rivers which drain the plateau of Central India. Floods have been occurring in all these rivers during the monsoon months of each year but it is observed that compared to north and northeastern rivers of the Indian sub-continent, flood magnitudes and frequencies are generally speaking far less in the peninsular rivers. It is seen from Figure 6 that there are only 6 G/D sites on 5 rivers, which have experienced more than 10 floods. Possibly, this is due to less number of cyclonic disturbances that visit this region during each monsoon period compared to northern Indian region. Also most of the river basins have their origin on the lee side of the Western Ghats, which experiences less rainfall activity as they happen to be in the rain-shadow zone of the Western Ghats and hence rainfall activity is not that severe over this region, unless the region comes under the influence of cyclonic disturbances.

It has also been seen that in Central Indian region, very large floods have occurred due to succession of cyclonic disturbances moving one after another within a short interval of time. In the case of west flowing rivers of Narmada and Tapi of Central India (see Figure 2), the course of the rivers and the track of monsoon disturbances, originating from the Bay of Bengal, mostly run parallel to each other. As a consequence of this, the flood wave from the upper reaches and heavy rainfall in the middle or lower reaches of these two rivers (viz., Narmada and Tapi) almost travel together resulting in severe floods in their lower reaches. This peculiarity is the speciality of the two west flowing rivers of the Narmada and the Tapi in Central India.

10. Most Severe Floods of the Country

A most severe flood has been defined as the one which is 5 meters and more above the danger level at a given G/D site on a river. Table IV gives a list of major rivers which recorded floods of 5 m and more above their respective D.Ls. It is seen from this table that 17 major rivers of this country recorded highest floods which were 5 m and more above their respective D.Ls at 31 G/D sites.

The regional distribution of these floods showed that although frequency of occurrence of floods is more in northeast India, but only 2 rivers of this region (viz., Barak and Manas rivers) recorded most severe floods of 5 m and above at their respective G/D sites. In north India only 9 rivers recorded 12 most severe floods at their respective G/D sites (see Table IV).

It is also seen from Table IV that 16 most severe floods were recorded in six central Indian rivers at 16 G/D sites. Among these six rivers, the Narmada river recorded the maximum number of (i.e., 7) most severe floods at 7 G/D sites. Next to this were the Subarnrekha and the Tapi Rivers, each of them recorded most severe floods at 3 G/D sites.

It may, however, be mentioned here, by way of information that the Teesta flood of 4th October, 1968 at Anderson Bridge in North Bengal is still considered to be the highest recorded flood in India which stood at 18.10 m above the D.L. at this site. Next to it is the Narmada flood of 6th September, 1970 at Gardeshwar in Gujarat which was 17.87 m above the D.L. These unprecedented or devastating floods were caused by the severe rainstorms of 3–5 October, 1968 and 5–7 September, 1970 respectively.

11. Worst Flood-Prone Gauge/Discharge (G/D) Sites on Different Rivers

From an examination of flood data of 15-years (1986–2000), it was observed that some of the rivers of north Indian region, viz. Adhwara Group, Bagmati, Burhi Gandak, Ganga, Ghagra, Kosi, Rapti, Sai and Sone, experienced floods of 2-day duration (or even more) at a time. As an example, in August 1988, the Ganga at Farakka recorded a flood which was 1.92 m above the D.L. for six consecutive days. Probably, this happened because these rivers flow through the plain areas where gradients are not so steep as in other regions and rainfall continues to occur for a longer duration of time.

It was also observed that only 11 rivers experienced floods during each and every monsoon season of the 15-year period. Table V gives a list of all these 11 rivers and number of floods experienced by them during each of the monsoon seasons. It is seen from Table V that during the 15-years period, Dibrugarh (on Brahmaputra) and Road Bridge (over Beki) are the two worst flood-hit G/D sites of the country. In the monsoon season of 1990, each of these two sites recorded 19 and 18 floods respectively, which are the highest number of floods in a year that any G/D site has recorded so far. It is also observed that July, August and September are the worst flood months for both these sites.

Table IV. Most severe floods of the country when flood levels were 5 m and more above their respective D.Ls. (updated upto 2000)

Sr. No.	River	G/D site	State	Danger level (m)	Deviation of highest flood from D.L. (m)	Date and year of occurrence
(A) Northeast India:						
1.	Barak	Lakhipur	Assam and	23.88	6.05	12.09.1979
2.	Manas	N.H. Crossing	neighbouring states	47.56	11.03	13.07.1984
(B) North India:						
3.	Chambal	Dholpur	Rajasthan	144.00	14.21	25.08.1982
4.	Ganga	Ballia	Uttar Pradesh	57.62	6.64	09.1996
5.	Jaldhaka	Bagarkata	West Bengal	185.60	7.75	1954
6.	Ken	Banda	Uttar Pradesh	104.00	9.24	14.09.1992
7.	Kosi	Barakshetra	Bihar	120.93	7.14	08.09.1973
8.	Mahi	Dam Axis	Gujarat	94.75	11.80	08.09.1973
		Vasaol	Gujarat	12.79	12.79	08.09.1973
9.	Ravi	Muktesar	Punjab	398.80	6.64	05.10.1965
10.	Teesta	Anderson Bridge	West Bengal	210.50	18.10	04.10.1968
		Coronation Bridge	West Bengal	149.40	10.65	10.1968
11.	Yamuna	Auraiya	Uttar Pradesh	108.96	6.63	30.07.1986
		Kalpi	Uttar Pradesh	105.00	7.70	27.08.1982
(C) Central India:						
12.	Baitarni	Dasarathpur	Orissa	38.36	9.32	16.08.1960
13.	Brahmini	Ponposh	Orissa	7.92	5.19	1971
14.	Godavari	Kanavaram	Andhra Pradesh	40.32	7.78	25.08.1990
15.	Narmada	Jamtara	Madhya Pradesh	374.90	5.69	1973
		Burmaghat	Madhya Pradesh	323.30	8.02	1973
		Hoshangabad	Madhya Pradesh	296.46	5.38	30.08.1973
		Mortakka	Madhya Pradesh	163.98	9.74	17.09.1961
		Dharamrai	Madhya Pradesh	122.83	6.17	01.09.1973
		Garudeshwar	Gujarat	23.77	17.87	06.09.1970
		Broach	Gujarat	7.31	5.94	07.09.1970
16.	Subarnrekha	Jamshedpur	Bihar	6.00	8.72	12.10.1973
		Jamsalaghat	Orissa	34.14	7.62	13.10.1973
		Rajghat	Orissa	10.36	7.01	1968
		Savangkheda	Maharashtrra	119.74	7.60	16.09.1959
		Gidhade	Gujarat	137.77	8.53	1968

Table V. List of G/D sites which experienced floods during each monsoon season of the period 1986-2000

Sr. No.	River	G/D site	No. of floods during the monsoon season of the year															Total
			1986	1987	1988	1989	1990	1991	1992	1993	1994	1995	1996	1997	1998	1999	2000	
(A) Northeast India:																		
1.	Brahmaputra	Dibrugarh	8	14	13	11	19	17	5	15	5	11	6	5	15	13	15	172
		Nematighat	8	14	13	9	12	10	5	11	4	11	6	3	14	6	4	130
		Tezpur	2	8	7	1	6	9	1	7	3	6	4	4	10	5	5	75
2.	Beki	Road Bridge	7	12	9	11	18	16	12	15	10	16	12	14	13	12	14	191
3.	Manas	N.H. Crossing	4	8	8	9	9	14	4	9	2	14	5	1	11	3	6	107
4.	Burhi-dihing	Khowang	2	9	12	9	12	11	2	3	2	8	3	2	10	5	2	92
5.	Dhansiri	Numaligarh	10	14	9	9	13	12	6	13	8	11	3	8	9	13	12	150
6.	Katakal	Matizuri	1	2	5	5	4	6	3	8	1	3	3	4	1	3	7	56
(B) North India:																		
7.	Ganga	Farakka	6	9	7	4	10	6	4	6	10	6	8	10	10	12	9	118
8.	Ghagra	Turtipar	7	2	9	6	7	5	2	6	4	5	6	10	7	12		92
9.	Kamla Balan	Jhanjarpur	5	9	10	9	11	4	1	6	6	8	10	7	11	7	4	108
10.	Kosi	Baltara	7	11	11	13	9	8	5	9	9	13	13	14	12	16	14	164
11.	Punpun	Sripalpur	7	8	2	2	8	3	1	4	5	3	4	8	3	6	5	69

From the above, it can, therefore, be said that as Cherrapunji and Mawsynram stations (Dhar and Nandargi, 1996) in northeast India happen to be the rainiest stations of the country, in the same way these two G/D sites, viz. Road Bridge and Dibrugarh, may be considered as the highest flood affected sites of the country.

12. Concluding Remarks

From the foregoing discussion it can be said that India's flood problem is mainly confined to the States located in the Indo-Gangetic plains and northeast India, where remedial measures must be taken to save these states from the yearly recurring ravages of devastating floods.

At present, the enormous water resources of the Brahmaputra river are not fully utilized and practically most of its flood water is running waste into the Bay of Bengal through Bangladesh after causing enormous damage by way of ravaging floods in Assam (in India) and Bangladesh. Though there is not much scope for increasing irrigation facilities in the Assam valley beyond a certain limit due to lack of proper storage sites especially in the main river and some of its tributaries, there is enormous potential for hydropower generation in this region especially at the "Great Bend" of the Dihang river when it enters India from Tibet and in some of its tributaries. All this has to be tapped fully if the northeast region as well as the country at large has to advance in the coming years so far power generation and agriculture are concerned. This, of course, requires full co-operation of the neighbouring countries of Tibet, Bhutan and Bangladesh for the development of water resources of this entire region.

Floods in the Ganga can be controlled to a large extent if the floods in the Himalayan rivers of the Ghagra, the Gandak and the Kosi can be controlled by withholding flood waters in the detention reservoirs built at sites in the Himalayas or where these rivers debauch into the Indian plains. The flood waters along with the heavy silt load cause flooding in the main Ganga during the monsoon months in the states of Uttar Pradesh, Bihar and West Bengal.

To utilize the enormous volume of flood water from the Himalayan rivers, it is suggested that multi-purpose water resources projects should be undertaken to conserve the flood waters and utilize them for agriculture, industry and hydro-power generation. This can be achieved only by building multi-purpose dams across these rivers within the Himalayas or at suitable sites where they debauch into the plains. In this respect the full co-operation and participation of the Nepal Government with the neighbouring Indian States is considered to be essential.

From the above it is clear that while the Ganga river system has abundant irrigation potentialities, similarly the Brahmaputra river has the highest hydro-electric potential. Very little of this potential has been made use of in this country so far. It is hoped that the present drought condition prevailing in some parts of India, would stimulate the authorities concerned and something tangible will be done to utilize the vast water resources available to this country from these river systems.

Acknowledgements

Authors are thankful to the Director, Flood Forecasting and Monitoring Directorate, Central Water Commission (CWC), New Delhi for the regular supply of Weekly Flood News Letters during the last few decades. Thanks are also due to India Meteorological Department, (IMD), Pune-5 for the supply of Meteorological Bulletins and relevant meteorological data associated with monsoon floods. Authors are grateful to the Dr. G. B. Pant, Director, and Dr. K. Rupa Kumar, Dy. Director, IITM, Pune-8 for giving all the necessary facilities to undertake this study.

References

Central Water Commission: 1986–2000, Weekly Flood News Letters, Flood Control Coordination Directorate (FCCD), New Delhi.

Chow, V. T.: 1956, Hydrologic studies of floods in the United States, *International Association of Hydrological Sciences* **42**, 134–170.

Das, P. K.: 1999, Vagaries of the summer monsoon rains, *Palaeobotanist* **48**, 1–6.

Dhar, O. N.: 1959, Meteorological situations associated with actual major floods in different parts of India, In: *Proc. Symp. on Meteorological and Hydrological Aspects of Floods and Droughts in India*, India Met. Dept. Publ.

Dhar, O. N. and Changrani, T. G.: 1966: A study of meteorological situations associated with major floods in Assam during the monsoon months, *Indian J. Meteorol. Geophys.* **17**(Spl. No.), 111–118.

Dhar, O. N., Bhattacharya, B. K., and Ghose, G. C.: 1975, A catalogue of the highest ever recorded floods in Indian rivers – a preliminary appraisal, *Indian J. Power & River Valley Development* **25**(12), 402–409.

Dhar, O. N., Rakhecha, P. R., and Mandal, B. N.: 1981a, Some facts about Indian rainfall – a brief appraisal from hydrological considerations, *Indian J. Power & River Valley Development* **31**(7&8), 117–125.

Dhar, O. N., Mandal, B. N., and Ghose, G.C.: 1981b, The Vamsadhara flash flood of September,1980 – a brief appraisal, *Vayu Mandal* **11**(3&4), 7–11.

Dhar, O. N., Rakhecha, P. R., Mandal, B. N., and Sangam, R. B.: 1981c, The rainstorm which caused Morvi dam disaster in August, 1979, *Hydrological Sciences Bulletin* **26**(1), 71–81.

Dhar, O. N., Ghose, G. C., and Kulkarni, A. K.: 1981d, A catalogue of major and devastating floods of India and their space-time distribution, In: *Proceedings of the International Conference on Flood Disasters*, Vol. I, Indian National Science Academy, New Delhi.

Dhar, O. N., Mandal, B. N., and Rakhecha, P. R.: 1984a, Rainfall distribution over India during the monsoon months in the absence of depressions and cyclonic storms, *Mausam* **35**(3), 309–314.

Dhar, O. N., Soman, M. K., and Mulye, S. S.: 1984b, Rainfall over the southern slopes of the Himalayas and the adjoining plains during 'Breaks' in the monsoon, *Journal of Climatology* **4**(6), 671–676.

Dhar, O. N., Mulye, S. S., and Mandal, B. N.: 1986, The highest ever recorded floods in the major Indian rivers – a brief appraisal. *Transactions of Institute of Indian Geographers* **8**(1), 13–26.

Dhar, O. N. and Nandargi, S.: 1992a, A study of rainspells associated with worst floods in the Jhelum river, *Journal of Applied Hydrology* **5**(1–4), 45–53.

Dhar, O. N. and Nandargi, S.: 1992b, A study of rainfall and floods in the 1987–1991 monsoon seasons over the contiguous Indian region, *Journal of Meteorology* **17**(174), 330–336.

Dhar, O. N. and Nandargi, S.: 1993, Worst flood-prone rivers and sites of India, *Vayu Mandal* **23**(3–4), 86–92.

Dhar, O. N. and Nandargi, S.: 1994, Floods in Indian rivers, *Indian J. Power & River Valley Development* **44**(7&8), 228–236.

Dhar, O. N. and Nandargi, S.: 1996, Which is the rainiest station in India, Cherrapunji or Mawsynram?, *Weather* **51**(9), 314–315.

Dhar, O. N. and Nandargi, S.: 1998, Floods in Indian rivers and their meteorological aspects, In: V. S. Kale (ed.), *Flood Studies in India* (Memoir No. 41), Geological Society of India, Bangalore.

Dhar, O. N. and Nandargi, S.: 2000, A study of floods in the Brahmaputra basin in India, *Inter. J. Climatol.* **20**(7), 771–781.

Framji, K. K. and Garg, B. C.: 1976, *Flood Control in the World – A Global Review*, Part I, International Commission on Irrigation and Drainage, New Delhi.

India Meteorological Department (IMD): 1986–2000, Weekly Weather Reports, DDGM (Forecasting), IMD, Pune.

Kale, V. S.: 1996, Introduction to Flood Hydrology, In: V. S. Kale (ed.), *Fluvial Geomorphology*, Department of Geography, University of Pune, Pune, India.

Kale, V. S.: 1998, Monsoon floods in India – a hydro-geomorphic perspective, In: V. S. Kale (ed.), *Flood Studies in India* (Memoir No. 41), Geological Society of India, Bangalore.

Kripalani, R. H. and Kulkarni, A.: 1997, Climatic impact of El-Nino/La Nina on the Indian monsoon – a new perspective, *Weather* **52**(2), 39–46.

Kulshrestha, S. M.: 1997, *Drought Management in India and Potential Contribution of Climate Prediction*, Joint COLA/CARE Technical Report No. 1, Centre for Ocean-Land-Atmosphere Studies/Centre for the Application of Research on the Environment, Maryland, 105 pp.

Kulshrestha, S. M.: 2000, *Flood Management in India*, Joint COLA/CARE Technical Report No. 3, Centre for Ocean-Land-Atmosphere Studies/Centre for the Application of Research on the Environment, Maryland.

Nandargi, S.: 1996, *Rainstorm Studies for Planning and Development of Water Resources of the Indian Region*, Unpublished Ph.D. Thesis, Pune University.

Nandargi, S. and Dhar, O. N.: 1998, An appraisal of successive tropical disturbances and their associated severe rainstorms during monsoon months, *J. Meteorol.* **23**(231), 221–228.

Pisharoty, P. R. and Asnani, G. C.: 1957, Rainfall around monsoon depressions over India, *Indian J. Meteorol. Geophys.* **8**(1), 1–6.

Rama Sastry, A.A.: 1984, *Weather and Weather Forecasting*, Publication Division, Government of India, New Delhi, India.

Sikka, D. R.: 2000, *Monsoon Floods in India*, Joint COLA/CARE Technical Report No. 4.

Centre for Ocean-Land-Atmosphere studies/Centre for the Application of Research on the Environment, Maryland, USA.

Ward, R. C.: 1978, *Floods – A Geographical Perspective*, The Macmillan Press Ltd., London.

WMO (World Meteorological Organization): 1974, *WMO/UNESCO International Glossary of Hydrology*, WMO Publication No. 385, WMO, Geneva.

WMO (World Meteorological Organization): 1994, *On the Front-Line: Public Weather Service*, WMO No. 816, WMO, Geneva, 12 pp.

Natural Hazards **28:** 35–64, 2003.

Three Recent Extreme Floods in Bangladesh: A Hydro-Meteorological Analysis

M. MONIRUL QADER MIRZA
Adaptation and Impacts Research Group (AIRG), The Institute for Environmental Studies (IES), University of Toronto, 33 Willcocks Street, Toronto ON M5S 3E8, Canada (E-mail: monirul.mirza@utoronto.ca also: mqmirza@hotmail.com)

(Received: 15 December 2001; accepted 4 March 2002)

Abstract. Bangladesh is highly vulnerable to floods due to its geographical location at the deltas of the Ganges, Brahmaputra and Meghna (GBM) rivers. About 92.5 per cent of the area of three basins lies outside the boundaries of the country. More than 80 per cent of the annual precipitation of Bangladesh occurs in the monsoon period between June and September. The hydro-meteorological characteristics of the three river basins are unique and they often cause large to extremely large floods in Bangladesh. It is possible that these floods could inundate 70 per cent of the country and the physical damage could be very serious for the economy of Bangladesh with its low gross domestic product (GDP). In 1987, 1988 and 1998, Bangladesh experienced three extreme floods, leaving trails of devastation and human misery. In this article it is demonstrated that these floods differed in terms of magnitude, extent, depth and duration. The external and internal hydro-meteorological dynamics were also different.

Key words: extreme floods, Bangladesh, Ganges, Brahmaputra and Meghna, hydro-meteorological characteristics and ENSO.

1. Introduction

Normal floods are important in the life of the people and economy of Bangladesh. People in rural and urban areas of Bangladesh have adapted to the changes introduced by normal floods, and are highly resilient. However, this resiliency is in jeopardy if an extreme flood event occurs. Such events are due to the unusual behaviour of the monsoon precipitation regime over the GBM river basins, compounded by drainage dynamics usually controlled by tidal inflow and backwater effects.

Bangladesh's location in the GBM basins is unique (Figure 1). The country contains only 7.5 per cent of the three river basins, while the remaining 92.5 per cent is located in China, Nepal, India and Bhutan (Mirza, 1997). The precipitation regimes of the three basins are significantly different, varying from low to very high areas of precipitation (Ahmed and Mirza, 2000). Very heavy precipitation during the monsoon in the upstream area generates a huge amount of runoff that flows through Bangladesh and drains to the Bay of Bengal through a narrow passage. Therefore, over bank spilling is a common occurrence causing floods in Bangladesh, where

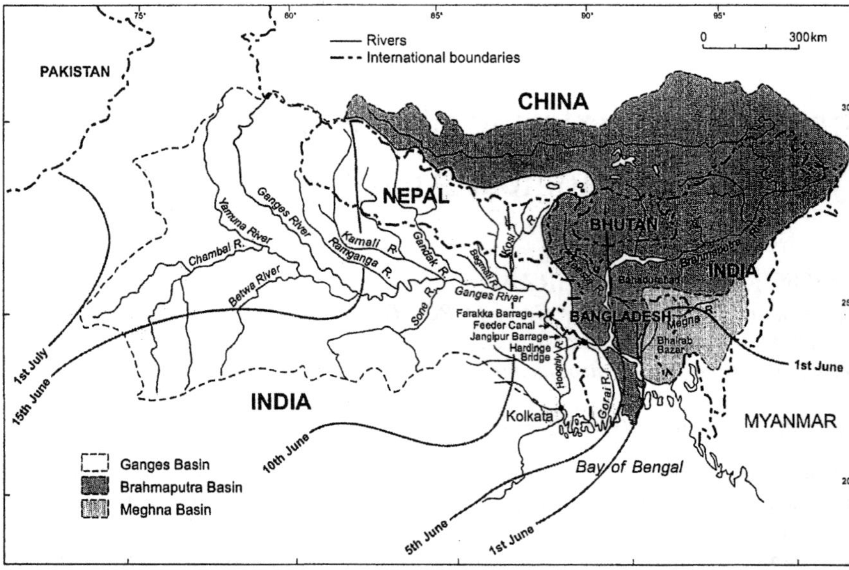

Figure 1. The Ganges, Brahmaputra and Meghna basins. Dates of onset of monsoon are also shown.

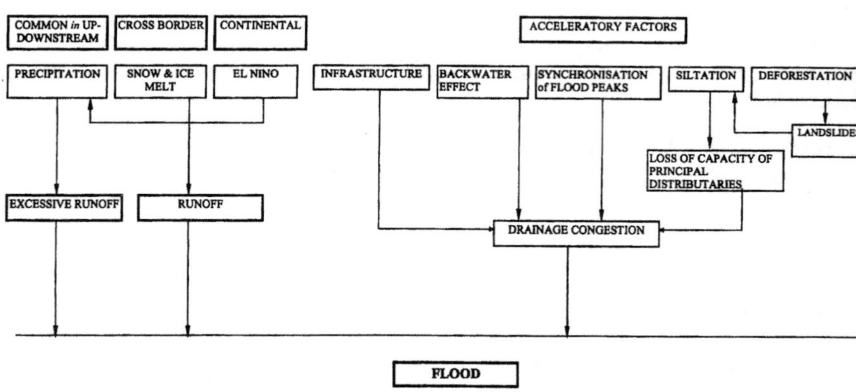

Figure 2. Causes of floods in Bangladesh.

local precipitation within the country aggravates the situation. In addition to intense monsoon precipitation, other factors that may cause floods include: snow and glacier melt, El Nino Southern Oscillation (ENSO) induced conditions, loss of drainage capacity due to the siltation of principal distributaries, backwater effect, unplanned infrastructure development, deforestation and the synchronisation of flood peaks of the major rivers (Ahmed and Mirza, 2000) (Figure 2).

Extreme floods in 1987, 1988 and 1998 engulfed as much as 70 per cent of Bangladesh (Figure 3). Economic damage and human misery, especially in rural areas, was enormous. These floods generated local and international debates on flood control in Bangladesh (Boyce, 1990). Thirteen donor countries formed a

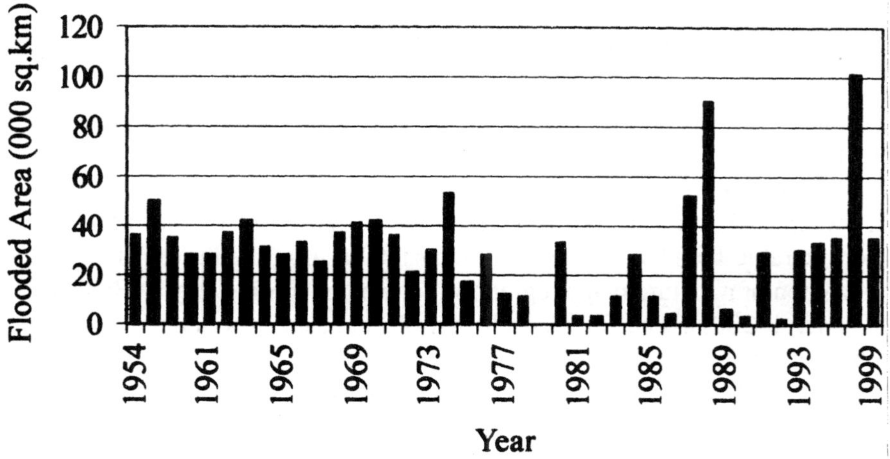

Figure 3. Area (× 1000 sq.km) flooded in various years in Bangladesh.

consortium which funded 27 studies under the 'Flood Action Plan (FAP)' (Mirza, 1991). Some analyses of the meteorological aspects of the 1987 and 1988 floods were carried out (UNDP, 1989; Brammer, 1990; Hofer, 1998) but contributions of external precipitation regimes were not adequately analysed. Detailed meteorological analysis of the recent 1998 flood in Bangladesh did not appear in the scientific literature, and a comparative analysis of the hydrological aspects of these three extreme floods did not get proper attention. Therefore, to fill the knowledge gaps, an attempt is made in this article to examine the external and internal precipitation regimes associated with the three extreme floods; and the hydrological aspects of the floods in terms of magnitude, depth and duration.

2. Monsoon Precipitation in the GBM Basins

Monsoons have wide ranging effects on life, society, hydrology and water resources, and the economy of major river basins in South Asia. On a long-term basis, there is a remarkable degree of regularity in the occurrence of the monsoon, but there are large variations in the amount of precipitation in time and space (Rogers *et al.*, 1989). These variations may cause either floods or droughts in the river basins. The onset of the Indian monsoon in terms of rainfall occurs over the Bay of Bengal in late May, and then over India about 1 June (Figure 1). The retreat of the monsoon from India occurs typically during September.

The first spell of the monsoon rain occurs around June 1 in the *Meghna basin* over north-eastern India and Bangladesh. Among the three river basins, the average precipitation (2600–3700 mm) of the Meghna basin is much higher than the other two, attributed to the topographical effects of the north-eastern part of India. For example, some of the world's highest precipitation totals have been recorded at Cherrapunji on the southern slopes of the Meghalaya plateau, where the moist air from the Bay of Bengal suddenly rises to 1,200 m (Rao, 1981). The copious precipitation in the Meghna basin in India has remarkable consequences for flash floods in the north-eastern region of Bangladesh, but local precipitation also contributes to this process. The monsoon completely retreats from the Meghna basin by 15 October.

The moist air mass of a monsoon arrives in the *Brahmaputra basin* in India between 1–5 June in a typical year. Before the onset of the monsoon, there is considerable thunderstorm activity over the basin in the month of May due to the incursion of moisture in the region from the neighbouring Bay of Bengal (Dhar and Nandargi, 2001). The Bangladesh part of the basin experiences monsoon rainfall about a week later, between 5–10 June. Average precipitation in the basin area in India and Bangladesh is 2350 mm and 1900 mm, respectively. In Bhutan, the average monsoon precipitation varies from less than 400 mm in the high Himalayas to 2000–4000 mm in the southern foothills. The monsoon retreats from the basin between 1–15 October.

Due to the vastness of the basin, the monsoon arrivals in the *Ganges basin* in India range between 5 June–1 July. The first spell of monsoon rain occurs by 5 June around Kolkata in India and in the adjoining area in Bangladesh. In the northern part of the basin area around Delhi, monsoon rain occurs around 1 July. The average monsoon precipitation in the Ganges basin in India varies from 360 mm on the western end to about 800 mm in the central area and to 1200–1600 mm near the delta (Mirza, 1997). In Nepal, monsoon precipitation is about 1500 mm. Rainfall in the *Terai region* of Nepal in the south along the Indian border is higher than the northern region. The monsoon retreats from north to south of the basin from 1 September–15 October.

3. Characteristics of Flood Flows in Bangladesh

The flood flows of Bangladesh rivers are unique in terms of timing of occurrence, magnitude, duration and severity. They are highly influenced by the precipitation patterns of the drainage areas outside the border and within Bangladesh.

3.1. TIME OF PEAK FLOW

Rivers in Bangladesh peak in different months during the monsoon in response to onset dates and precipitation characteristics. The flood hazard in the central region (around the Baruria Transit) can increase significantly if floods arrive simultaneously in the Ganges and Brahmaputra rivers. Assessment of the timing of the river peaks is thus important in order to provide estimates of the magnitudes of discharge, flooded areas and damage. An analysis of occurrences of floods in some major rivers and their tributaries was made so that the months with greatest potential for generating simultaneous floods could be identified. For the analysis, monthly maximum discharges for various periods were compiled from FEC (1989)

Table I. Occurrence of maximum monthly peak discharge (in %)

River/Station	June	July	August	September	Number of peak years
The Brahmaputra family Brahmaputra (Bahadurabad)	9	41.0	35.0	15.0	34.0
Teesta (Kaunia)	0	58.0	37.5	4.5	24.0
Old Brahmaputra (Mymensingh)	0	12.5	75.0	12.5	8.0
The Ganges family Ganges (Hardinge Bridge)	0	0	45.0	55.0	53.0
Padma (Baruria)	0	19.0	57.0	24.0	21.0
Brahmaputra family*	5	44.0	41.0	10.0	
Weighted Average	0	5.0	48.0	47.0	

Denotes weighted average (in terms of occurrence) distribution. Indented rivers are tributary and distributary.

and BWDB (1995). Table I Shows that the peaks in the Brahmaputra family* of rivers occur in July and August and in the Ganges family of rivers, in August and September. Maximum discharge in the Brahmaputra occurs 35 per cent of the time and for the Ganges occurs 45 per cent of the time in August. This indicates a real likelihood of simultaneous floods in the Ganges and Brahmaputra rivers in August. The weighted average of these two families of rivers also indicates a high possibility of flooding in August. The similarity in the time of occurrence of peaks in the Brahmaputra and Teesta rivers shows that they drain areas with similar climatic patterns.

3.2. DISTRIBUTION OF LAG-TIMES OF ANNUAL PEAKS

An assessment of lag-times between the yearly peaks of the two major rivers helps determine the possibility of simultaneous flood peaks. In this regard, a 28-year period common to Bahadurabad and Hardinge Bridge was analysed and presented in Table II showing that the occurrence of "simultaneous" (assumed to occur within one day period) peaks in the Ganges and Brahmaputra rivers is not a rare event. On average, the Brahmaputra peaks occur 29 days earlier than the Ganges. However, the lag-times are well scattered. The distribution shows one group (0–10 days) that has occurred 22 per cent of the time over a 28-year period.

* FEC (1989) defined the family of rivers according to the peaking time of discharge.

Table II. Distribution of lag-times between yearly peaks of the Ganges River at Hardinge Bridge and Brahmaputra River at Bahadurabad

Time-Lag(days)	Number (cumulative)	Percentage distribution	Percentage (Cumulative)
−82 to −72	–	3.57	–
−71 to −24	2	7.14	7.14
−23 to −14	3	3.57	10.71
−13 to −1	4	3.57	14.28
0–10	10	21.43	35.71
11–30	12	7.14	42.85
31–40	18	21.43	64.28
41–50	21	10.72	75.00
51–60	25	14.28	89.28
61–70	26	3.57	92.85
71–88	28	7.15	100.00

A minus sign indicates the Ganges peaks occurred first. The shaded area indicates "period of simultaneous occurrence of floods.

3.3. CRITICAL PEAK PERIODS FOR THE GANGES AND BRAHMAPUTRA RIVERS

The next important issue is the identification of peak periods for the Ganges and Brahmaputra rivers. From records of the peaks (a sample of 28 years was considered), two specific periods have been identified for the Brahmaputra.

3.3.1. *Period One*

A first flood occurs in the Brahmaputra basin at around day 128 (August 7) of the water year (1 April to 31 March) with the normal occurrence of the Brahmaputra peak. In response to the precipitation and basin characteristics, the Brahmaputra River generates a peak in advance of the Ganges River.

3.3.2. *Period Two*

The late Brahmaputra peak arrives at around day 153 (August 31) of the water year, which overlaps with the average Ganges peak that occurs at around day 157 (September 4). In some years, the annual peak occurred during this second period. During period two, the likelihood of a simultaneous occurrence is very high. Peaks in the Teesta and Meghna rivers occur simultaneously because they drain catchments with similar characteristics. This indicates the possibility of simultaneous flooding in all river basins is higher with the late Brahmaputra peak. In a given year, the Brahmaputra may experience two flood peaks. For example, a second peak may occur in August, with serious implication if it coincides with the normal Ganges peak.

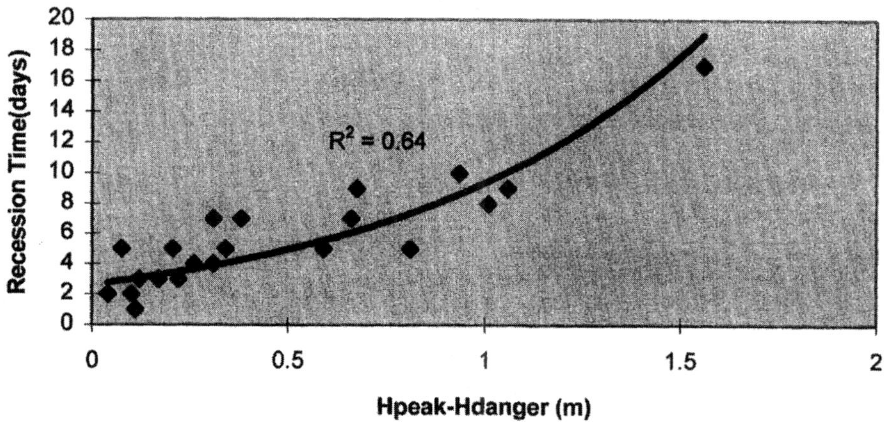

Figure 4. Relationship between depth of water above danger level and recession time for the Brahmaputra River at Serajgonj. Hpeak = Peak water level (m + PWD) and Hdanger = Danger level (m + PWD).

3.4. DURATION OF FLOOD DISCHARGE

The duration of a flood is important in terms of damage and recovery, and is a function of inflow of discharge, drainage capacity of the river channel, backwater effect, tidal activity, etc. Although many of these factors have synergetic effects on flood duration, a simple relationship between depth of water above a critical level and recession time of various floods are plotted in Figure 4 for the Serajgonj station on the Brahmaputra River. With the increases in flood level, recession time increases exponentially. Other major rivers show similar behaviour.

3.5. SEVERITY OF PEAK DISCHARGE

Flood volume alone does not explain the severity of flood per unit area of the basin. The severity can be assessed by determining the regional coefficient '*K*', using the Francou-Rodier relationship (FEC, 1989). The coefficients are calculated using predicted peak flows of different return periods for several rivers and arranged in decreasing order Table III.

The *K* values (Table III) indicate that, per unit area, the severity of flooding in the Brahmaputra River basin is the highest of all the basins. The basin areas of the Teesta and Meghna are small, but severity of flooding is substantial because of the precipitation influence of Sub-Himalayan West Bengal and the Assam ranges. The *K* value for the Meghna shows the damping effect of the Meghna depression (FEC, 1989). The Padma is a hybrid of the Ganges and Brahmaputra rivers. Its K value is strongly influenced by the Brahmaputra River.

Table III. Francou-Rodier's regional coefficients

River/Station	Basin Area (10^3 sq.km)	K (10)	K (20)	K (25)	K (50)	K (75)	K (100)	Remark
Brahmaputra (Bahadurabad)	536.0	5.23	5.38	5.42	5.56	5.63	5.68	1
Padma (Baruria)	1490.0	4.79	4.97	5.02	5.19	5.28	5.33	3
Teesta (Kaunia)	12.3	4.51	4.65	4.69	4.80	4.86	4.90	1
Meghna (Bhairab Bazar)	60.7	4.54	4.65	4.75	4.78	4.82	4.85	1
Ganges (Hardinge Bridge)	907.0	4.14	4.32	4.37	4.52	4.61	4.67	2

1. Brahmaputra Regime. 2. Ganges Regime. 3. Brahmaputra-Ganges Regime. Francou-Rodier's Coefficient (K) is determined from the relationship $Q/10^6 = (A/10^8)$ *Exp $(1 - K/10)$; Q in m^3/sec and A in sq.km.

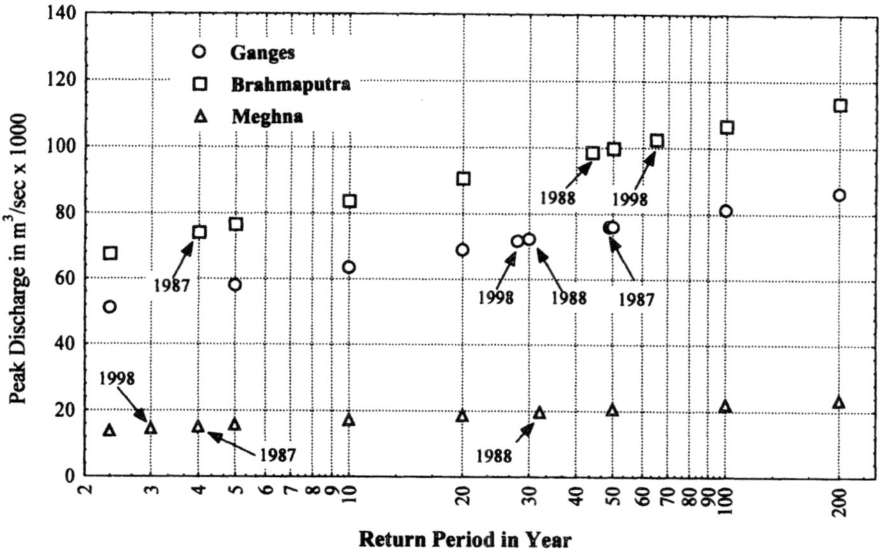

Figure 5. Return periods of peak discharges of the GBM rivers.

3.6. RETURN PERIODS OF PEAK DISCHARGE AND WATER LEVEL

The return period of an extreme event is inversely related to its frequency of occurrence, and very severe events occur less frequently than more moderate events. A frequency analysis was carried out to relate the magnitude of extreme events to their frequency of occurrence through the use of probability distributions. The analysis demonstrates how frequently these three events may occur. This analysis has substantial implications for design of water control structures and their economic values, delineation of flood plains and determining the effect of encroachments on the flood plains.

For flood frequency analyses, the Gumbel Type I distribution (EV1) has been recommended for the major rivers in Bangladesh (FAP, 1992). Accordingly, the EV1 distribution was applied for estimating return periods of peak discharge and water level for the Ganges, Brahmaputra and Meghna rivers and presented in Figure 5 and Figure 6. The EV1 probability distribution function is:

$$F(x) = \exp\left[-\exp\left(-\frac{x-u}{\alpha}\right)\right] \qquad -\infty \leq x \leq \infty \tag{1}$$

where $F(x)$ is the probability of an annual maximum $Q \leq x$ and α and u are scale parameter and location parameter, respectively. Defining s as standard deviation and x as mean, the expressions for α and u are as follows:

$$\alpha = \frac{\sqrt{6}s}{\pi} \tag{2}$$

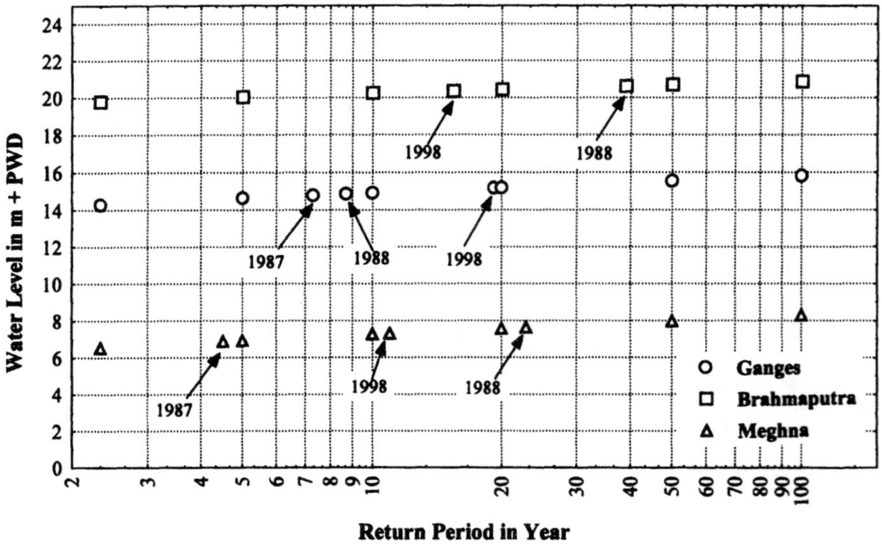

Figure 6. Return periods of peak water levels of the GBM rivers.

$$u = \bar{x} - 0.5772\alpha. \tag{3}$$

With a finite sample, the mean and standard deviation can be estimated from the moments of the data sample. The probability model can be obtained by substituting the value of α and u in Equation (1). Figure 5 and Figure 6 present the magnitude of discharges and water levels of the Ganges, Brahmaputra and Meghna rivers for various return periods. Due to the relatively high coefficients of variations of discharge and water level for the Ganges River, the magnitude of uncertainty for return periods is higher than that of the Brahmaputra and Meghna rivers.

4. The Extreme Floods of 1987, 1988 and 1998: Hydro-Meteorological Analysis

Three extreme floods devastated Bangladesh in the year 1987, 1988 and 1998, when 36, 63 and 69% area of the country was engulfed (Figure 3). The floods occurred in different months in response to high monsoon rainfall in Bangladesh (rainfall stations are shown in Figure 7) and cross-border basin areas. Floods in 1987 (Figure 8) began in late June when flash floods occurred in the north-eastern districts of Sylhet and Sunamgonj (BWDB, 1987) as a result of high rainfall. In late July, the Brahmaputra and Ganges exceeded their danger levels at Chilmari and Goalundo, respectively (Figure 7). The Meghna exceeded its danger level at Brahrab Bazar in early August. The floods of 1988 (Figure 9) began when with some of the rivers in the South-eastern hill basins crossing their danger levels in early May. The Brahmaputra River remained in flood stage from early July and had two other peaks prior to the catastrophic flood of August–September. The late extreme peak in the Brahmaputra coincided with the moderately high peak in the Ganges River with a time difference of only three days. The Meghna flood reached the highest level at the end of first week of September. The flood in 1998 (Figure 10) started in the late second week of July and continued till the middle of September and in some places the whole of September (Chowdhury, 2000). Districts in the northern and central parts and a part of South-west Bangladesh were inundated mainly due to the onrush of flood waters from India. The flood situation worsened in the three river basins in the first week of September. Magnitudes of the peak discharge of the major rivers were different during the three flood years. Flood depths were also different and floods remained differing periods in the upstream and downstream areas.

4.1. THE PRECIPITATION REGIMES

Occurrence of rainfall in the Indian subcontinent is correlated with El Nino Southern Oscillation (ENSO) (Miyakoda *et al.*, 2000). A study of the Indian monsoon has shown that there is a good correlation between the strong Southern Oscillation Index (SOI) (cold phase) or "La Nina" and good monsoon years (Parthasarathy and Pant, 1985). Similar research on Bangladesh monsoon rainfall (Ahmed, 1999) has shown that in general there is a decrease in rainfall in El Nino years (warm phase) in all three seasons - the pre-monsoon, monsoon and post- monsoon. The SOI is a measure of difference in pressure anomaly between Darwin in Australia and Tahiti Island situated in the eastern Pacific. When this index is positive, it reflects a cold phase of sea surface temperature in the southern Pacific Ocean. During an El Nino event (warm phase) this index is negative. During an El Nino period there is an abrupt rise of temperature in the eastern Pacific Ocean and as a result the atmospheric circulation in the region changes. The easterly wind from the Pacific to the Bay of Bengal, which brings a lot of moisture, reverses or weakens. As a result, Bangladesh and other parts of the Indian subcontinent receive less rain during an El Nino year. On the other hand, when the temperature in the eastern Pacific decreases, the easterly wind becomes strong, the flow of moisture is abundant and there is increased rainfall in Bangladesh and neighboring regions, resulting heavy flooding.

4.1.1. *1987 Precipitation Regime*

The year 1987 was an El Nino year (Figure 11). Krishnamurti *et al.* (1987) summarized the monthly meteorological conditions and sea surface temperature from April to August of 1987 on a global scale. They concluded that the planetary scale divergent circulation at 200 mb shifted eastward and equatorward, due to the El Nino related warm sea surface temperature over the equatorial Pacific, and that this eastward shift of the divergent circulation in the upper troposphere caused severe drought over most of the Indian subcontinent, and flooding in northeast India and Bangladesh. In India, the monsoon season's rainfall departure was -19% and

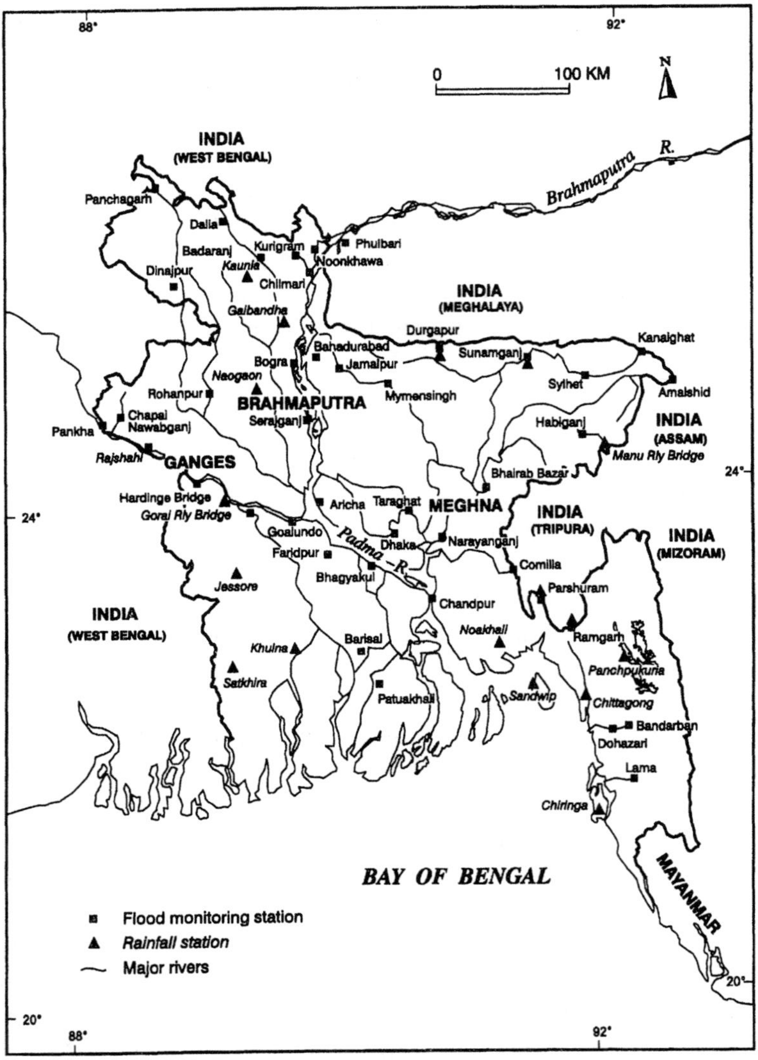

Figure 7. Flood monitoring and rainfall stations of the BWDB.

the area under deficient rainfall was 64% (Kulshrestha, 1997). Overall monsoon rainfall across Bangladesh was significantly above the long-term mean (Table IV and Figure 7).

In 1987, severe floods occurred in the Brahmaputra basin in eastern India and Ganges basin in Bihar due to a break monsoon,[2] one of the components of

[2] Rao (1976) defined the break monsoon as follows: the monsoon trough is located close to the foot of Himalayas which leads to a striking decrease of rainfall over most of India but an increase along the Himalayas and parts of northeast India and southern Peninsula. Dhar *et al.* (1984) showed that the increase in rainfall during break monsoon periods is at maximum at the southern foothill of Himalayas between longitudes 85° and 87° E around 27° N.

Table IV. Monthly rainfall (mm) in Bangladesh during 1987 flood

Basin/Station	July			August			September		
	Normal	1987	Deviation	Normal	1987	Deviation	Normal	1987	Deviation
Ganges Basin									
Panchagarh	836	1149	313	568	1238	670	276	809	533
Dinajpur	455	947	492	318	843	525	283	415	132
Rajshahi	334	405	71	241	424	183	221	235	14
Pabna	281	332	51	278	491	213	218	169	−49
Khustia	311	306	−5	292	491	199	200	115	−85
Jessore	329	533	204	309	751	442	239	159	−80
Khulna	370	570	200	301	666	365	249	168	−81
Faridpur	337	424	87	300	375	75	244	433	189
Barisal	481	433	−48	384	798	414	332	237	−95
Brahmaputra Basin									
Kurigram	455	870	415	297	607	310	296	311	15
Kaunia	504	1555	1051	318	644	326	341	270	−71
Rangpur	443	1378	935	329	525	196	280	396	116
Bogra	401	582	181	321	734	413	237	214	−23
Dewangonj	460	1062	602	318	535	217	244	425	181
Jamalpur	440	865	425	363	702	339	253	604	351
Mymensingh	460	408	−52	390	517	127	310	407	97
Tangail	283	336	53	286	356	70	211	341	130
Dhaka	360	541	181	333	464	131	234	340	106
Meghna Basin									
Sylhet	721	1290	569	599	735	136	439	953	514
Sunamgonj	1410	1629	219	1079	690	−389	596	1701	1105
Sheola	762	793	31	552	454	−98	413	781	368
Moulvi Bazar	452	584	132	391	490	99	269	209	−60
Habigonj	402	548	146	393	443	50	255	276	21
Durgapur	852	941	89	608	659	51	526	629	103
Bhairab Bazar	432	305	−127	325	474	149	266	485	219
Comilla	492	305	−187	369	445	76	263	357	94
Chandpur	380	771.4	391.4	320	527	207	235	251	16
South Eastern Hill Basin									
Parshuram	557	768	211	414	595	181	227	294	67
Noakhali	629	1287	658	595	836	241	419	696	277
Rangamati	573	783	210	465	902	437	260	425	165
Chittagong	680	1303	623	568	616	48	294	515	221
Panchpukuria	645	1189	544	462	668	206	234	487	253
Bandarban	612	647	35	424	796	372	270	353	83
Lama	777	1263	486	533	926	393	315	397	82

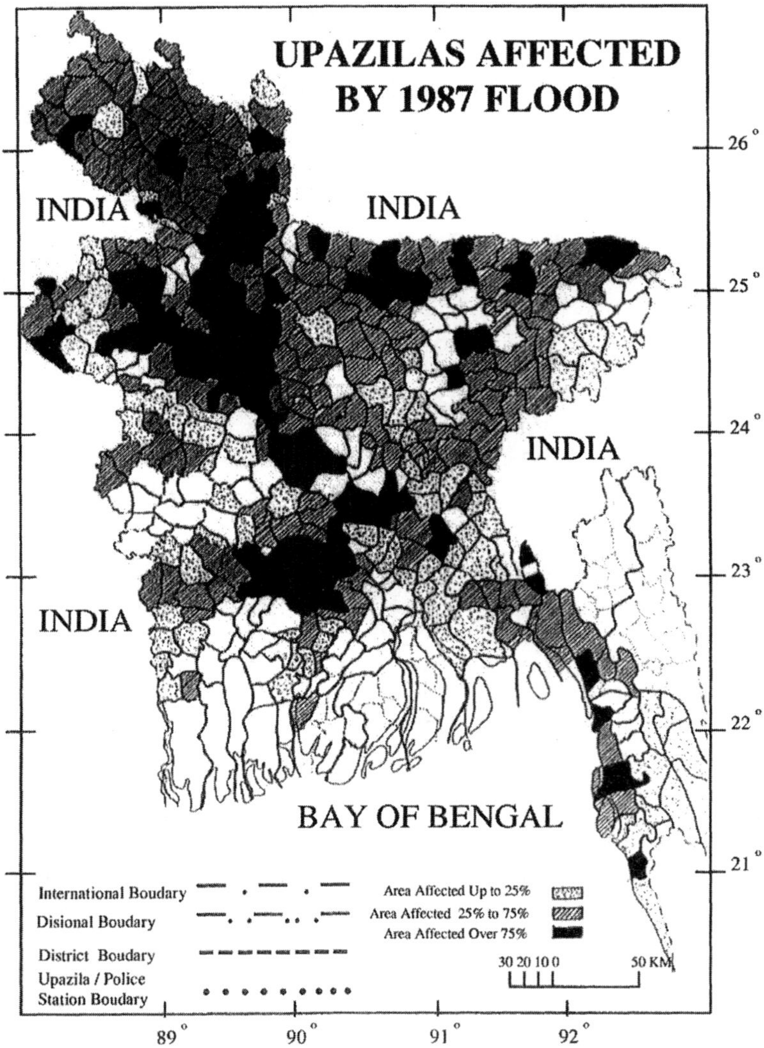

Figure 8. Extent of flood in 1987. Upazila denotes the lowest administrative district.

interseasonal variation of monsoon activity. These two regions are close to the Bangladesh border and parts of the shared river basins. In Assam in eastern India the area engulfed was the second highest since 1953 (3.08 million ha) (CSE, 1992). Precipitation in the bordering meteorological subdivisions of South Assam and Sub Himalayan West Bengal in the Bramhaputra basin was notably higher than normal. A similar precipitation pattern was noticed in the Gangetic West Bengal and Bihar plain meteorological subdivisions in the Ganges Basin (Table V and Figure 12). The cross-border flood waters generated from the excess precipitation eventually moved into Bangladesh.

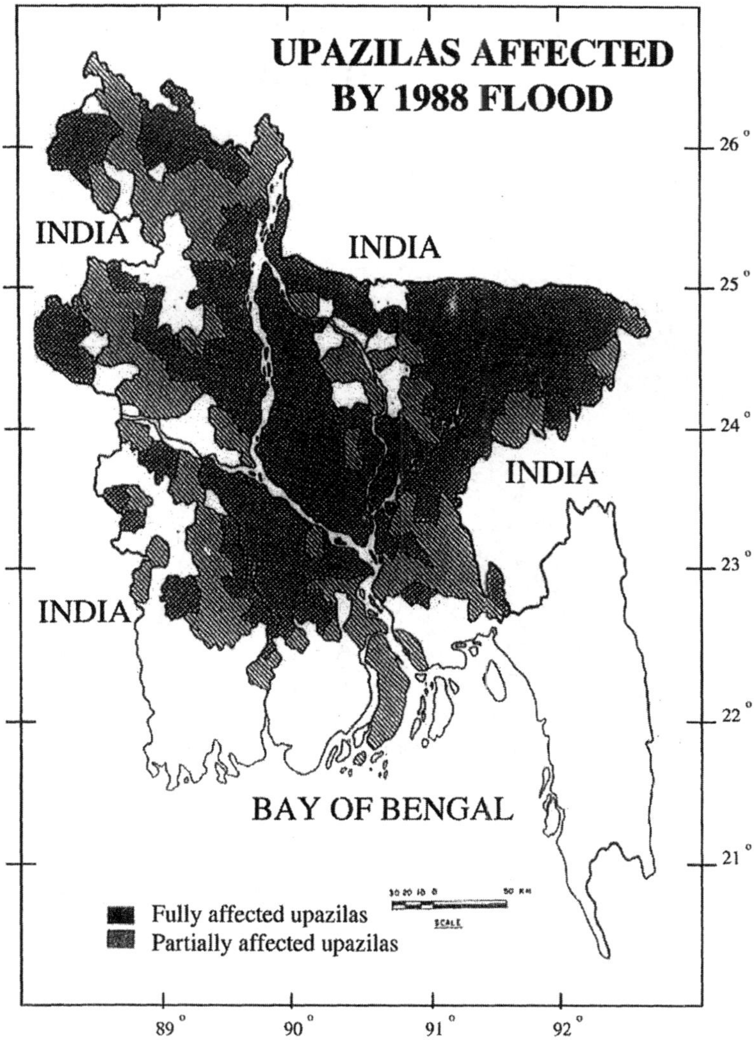

Figure 9. Extent of flood in 1988. Upazila denotes the lowest administrative district.

Before the peak time of flow of the main rivers in 1987, precipitation across the basin areas in Bangladesh filled all water bodies and low-lying areas. For example, in July rainfall in the four basins within Bangladesh was much higher than the normal (Table IV). Most noticeable was the 100% higher rainfall in the Brahmaputra basin than the long-term mean. The Ghagot and Teesta sub-basins of the Brahmaputra basin also experienced huge rainfall. The Ganges basin (downstream of the Hardinge Bridge) in Bangladesh experienced torrential rainfall in August, which was twice the normal (Table IV). In 1987, the Ganges attained two flood peaks – one in August and another in September. The Brahmaputra discharge peaked in August. The Ganges as well as the Brahmaputra peaks were not cor-

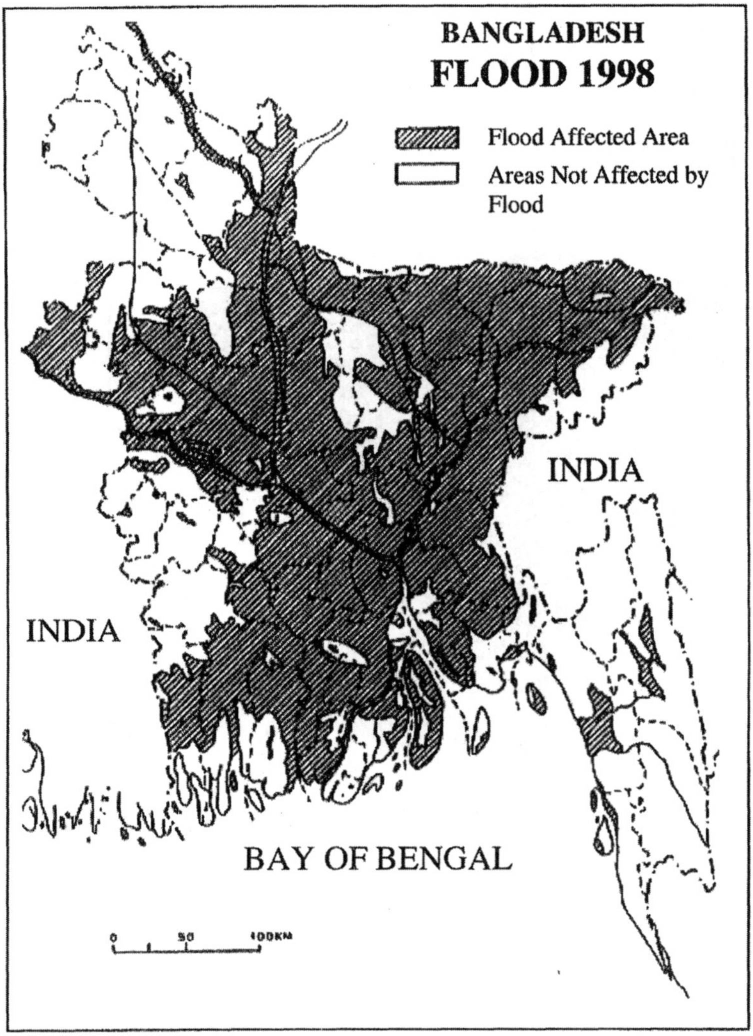

Figure 10. Extent of flood in 1998.

related with the rainfall processes in Bangladesh, thus only the rainfall situation outside Bangladesh can explain the situation (Hofer, 1998). The extraordinary peak of the Teesta from 10–16 August coincides with the peak of the Brahmaputra that caused floods in Northwestern Bangladesh. But the rainfall that occurred in the region was not sufficient to explain this situation. The Meghna reached its peak in the middle of August. Overall, local rainfall in Bangladesh throughout the monsoon months aggravated the flood situation in the four basins (Table IV).

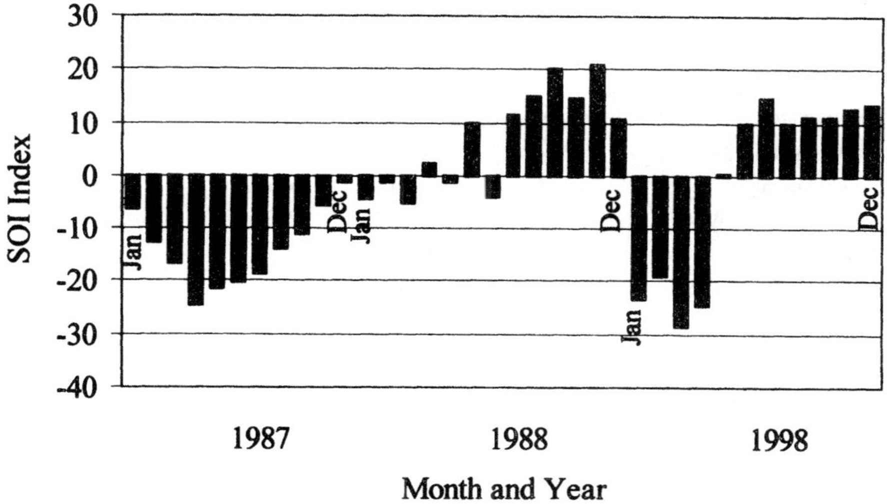

Figure 11. The southern oscillation index (SOI).

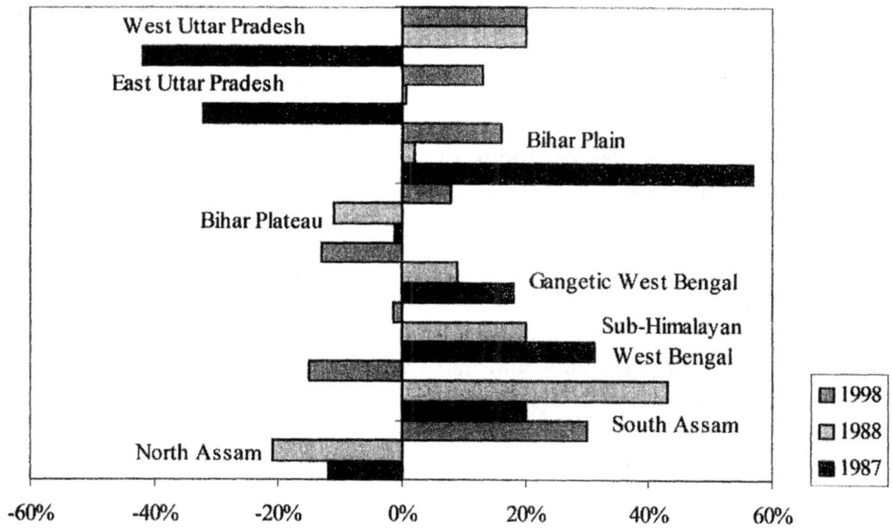

Figure 12. Deviations (%) of rainfall from normal in some meteorological subdivisions in the GBM basins in India.

4.1.2. *1988 Precipitation Regime*

In 1988, abundant rains occurred in the GBM basins area with the “La Niña” event. As seen from the SOI index in Figure 11 that from September 1987 onwards, the central equatorial pacific began retreating from the warming phase to the colder phase and entered into the colder phase in March 1988. By July the “La Niña” phase began and continued until December 1988.

Table V. Monsoon rainfall (mm) in various meteorological subdivisions of the GBM rivers in India

Meteorological subdivision	Normal	1987	Deviation	1988	Deviation	1998	Deviation
Brahmaputra and Meghna basins							
North Assam	1969	1738	−231	1546	−423	2553	584
South Assam	1247	1507	260	1787	540	1057	−190
Sub Himalayan West Bengal	2038	2665	627	2448	410	2017	−21
Ganges Basin							
Gangetic West Bengal	1097	1272	175	1196	99	958	−139
Bihar Plateau	1078	1067	−11	963	−115	1167	89
Bihar Plain	1005	1585	580	1029	24	1164	159
East Uttar Pradesh	898	607	−291	903	5	1019	121
West Uttar Pradesh	774	447	−327	935	161	929	155
Haryana	513	142	−371	786	273	729	216
West Rajasthan	243	98	−145	272	29	221	−22
East Rajasthan	630	361	−269	590	−40	628	−2
West Madhaya Pradesh	922	684	−238	863	−59	885	−37
East Madhaya Pradesh	1145	945	−200	945	−200	854	−291

Source: IMD (2001).

There were two waves of floods in 1988 in Bangladesh, which were mainly caused by runoff generated in the cross-border basin areas. During the *first wave* of flood in July a storm crossed the Meghalaya, entered the Assam Valley and caused extensive rainfall. During July 1–9, severe rainfall (2633 mm) was recorded in Cherrapunjee (Hofer, 1998). The Teesta, Dharala and Brahmaputra rivers in Bangladesh all exceeded crossed their danger levels. The Brahmaputra remained in flood stage from early July and had two flood peaks on 10 July and 30 July. However, the flood in the first part of July in the Meghna basin is clearly related to the rainy phase (mid June-mid July) that was recorded at almost all stations across the basin (Table VI). The annual peak occurred at Bhairab Bazar station on 11 July.

The *second wave* of flood hit Bangladesh in late August and early September. Some rainfall occurred in the northern and northwestern part of the country, but the rains were not adequate to cause such a huge flood (Hofer, 1988). Therefore, an important cross border input has to be assumed in addition to the high rainfall in some parts of Bangladesh itself. Very high rainfall was reported in the Sub Himalayan West Bengal and Gangetic West Bengal meteorological subdivisions in the latter part of August (GOI and GOB, 1990), leading to the peaking of the flow in the Ganges River at Hardinge Bridge on 3 September. Rainfall in the basin area in Bangladesh was also substantially higher (Table VI). In the 1988 flood, the Brahmaputra River at Bahadurabad reached its peak on August 30. During the week August 18–24 and August 25–31, precipitation levels in Aurunachal Pradesh, Assam and Meghalaya were on average 336% and 76% higher than normal (GOB and GOI, 1990). Precipitation in Sub-Himalayan West Bengal was 176% above normal, which contributed to the peak of the Teesta River (26 August, 1988). Therefore, in 1988, excess precipitation in the Brahmaputra basin in India contributed significantly to the flood peak at Bahadurabad, Bangladesh. Some contribution from Bangladesh was also observed. In August, rainfall in Kaunia and Rangpur was 130 and 84% higher than normal. Downstream of Bahadurabad, rainfall in Dewangonj was 30% and Jamalpur 36% higher than normal which contributed to flooding in the areas downstream of Bahadurabad. In August, high rainfall was recorded in the stations upstream of Bhairab Bazar in the Meghna basin (Table VI and Figure 7). In five out of eight BWDB stations used for flood forecasting, rainfall was 44–132% higher than average (Table VI). However, downstream of Bhairab Bazar, rainfall was slightly less than the normal. In September, rainfall in three upstream stations was 29–83% higher.

4.1.3. *1998 Precipitation Regime*

In 1998, a "La Nina" event similar to that of 1988 brought huge rains to the region and was the main cause of floods in Bangladesh (Chowdhury, 2000a, b) (Figure 11). In 1998, part of the Ganges basin in India experienced higher rainfall than normal in Bihar Plateau, Bihar Plain, East Uttar Pradesh and West Uttar Pradesh from late August to the middle of September (Table V and Figure 12). The external contribution to the flood discharge of the Ganges River in Bangladesh came from Bihar and Uttar Pradesh. Upstream of the Brahmaputra Basin in India, high rainfall was concentrated in the North Assam meteorological subdivision and a significant departure is noticeable (Figure 12). In 1998, Dhar and Nandargi (2001) reported 14 flooding events at Dhubri gauging site on the Brahmaputra River, which eventually reached downstream to Bangladesh.

In the main flooding period (late August–middle of September), rainfall in the Ganges Basin was generally normal except for three stations (Table VII). Three stations on the left bank of the Ganges River experienced lower than normal rainfall. Therefore, local contribution to flooding in the Ganges basin was negligible in 1998. In August, rainfall in the Brahmaputra basin in Bangladesh was also significant. In the Teesta basin, rainfall was 32–122% higher than normal, contributing to the peak discharge of the Teesta River, which was already high due to the external contribution from India. Peak discharge of the Teesta River merged with the inflow of the Brahmaputra River from India and generated 102,534 m^3/sec flood peak at Bahadurabad, the highest ever in the recorded history. Stations downstream of Bahadurabad also experienced high rainfall (42–97% higher the normal) contributing to flooding in the central region of Bangladesh (Table VII). The external contribution to the flood discharge of the Meghna River was not significant in 1998. In the main flooding period, local rainfall in the Meghna basin contributed significantly to flooding. The basin area upstream of Bhairab Bazar experienced rainfall 15–71% higher than normal (Table VII), raising the flood peak of the river at Bhairab Bazar.

Table VI. Monthly rainfall (mm) in Bangladesh during 1988 flood

Basin/Station	July			August			September		
	Normal	1988	Deviation	Normal	1988	Deviation	Normal	1988	Deviation
Ganges Basin									
Panchagarh	836	823	−13	568	1341	773	276	481	205
Dinajpur	455	489	34	318	893	575	283	198	−85
Rajshahi	334	343	9	241	363	122	221	84	−137
Pabna	281	320	39	278	181	−97	218	180	−38
Khustia	311	300	−11	292	274	−18	200	128	−72
Jessore	329	277	−52	309	422	113	239	94	−145
Khulna	370	343	−27	301	402	101	249	56	−193
Faridpur	337	227	−110	300	264	−36	244	190	−54
Barisal	481	429	−52	384	492	108	332	194	−138
Bramaputra Basin									
Kurigram	455	666	211	297	723	426	296	333	37
Kaunia	504	438	−66	318	730	412	341	391	50
Rangpur	443	474	31	329	608	279	280	523	243
Bogra	401	586	185	321	272	−49	237	281	44
Dewangonj	460	538	78	318	407	89	244	317	73
Jamalpur	440	376	−64	363	494	131	253	267	14
Mymensingh	460	657	197	390	399	9	310	406	96
Tangail	283	404	121	286	292	6	211	*	*
Dhaka	360	253	−107	333	168	−165	234	187	−47
Meghna Basin									
Sylhet	721	723	2	599	1152	553	439	604	165
Sunamgonj	1410	1726	316	1079	2514	1435	596	1093	497
Sheola	762	505	−257	552	809	257	413	621	208
Moulvi Bazar	452	386	−66	391	667	276	269	207	−62
Habigonj	402	584	182	393	567	174	255	331	76
Durgapur	852	920	68	608	937	329	526	502	−24
Bhairab Bazar	432	439	7	325	334	9	266	291	25
Comilla	492	640	148	369	346	−23	263	239	−24
Chandpur	380	332	−48	320	276	−44	235	156	−79
South Eastern Hill Basin									
Parshuram	557	907	350	414	471	57	227	344	117
Noakhali	629	526	−103	595	894	299	419	631	212
Rangamati	573	286	−287	465	521	56	260	*	*
Chittagong	680	798	118	568	414	−154	294	224	−70
Panchpukuria	645	632	−13	462	520	58	234	448	214
Bandarban	612	579	−33	424	420	−4	270	430	160
Lama	777	472	−305	533	438	−95	315	349	34

* Not available.

Table VII. Monthly rainfall (mm) in Bangladesh during 1998 flood

Basin/Station	July			August		
	Normal	1998	Deviation	Normal	1998	Deviation
Ganges Basin						
Panchagarh	836	1154	318	568	903	335
Dinajpur	455	686	231	318	555	237
Rajshahi	334	363	29	241	261	20
Pabna	281	422	141	278	275	−3
Khustia	311	429	118	292	241	−51
Jessore	329	213	−116	309	279	−30
Khulna	370	373	3	301	309	8
Faridpur	337	312	−25	300	351	51
Barisal	481	661	180	384	524	140
Brahmaputra Basin						
Kurigram	455	533	78	297	661	364
Kaunia	504	612	108	318	420	102
Rangpur	443	477	34	329	616	287
Bogra	401	745	344	321	504	183
Dewangonj	460	684	224	318	617	299
Jamalpur	440	439	−1	363	517	154
Mymensingh	460	796	336	390	767	377
Dhaka	360	515	155	333	562	229
Meghna Basin						
Sylhet	721	1347	626	599	779	180
Sunamgonj	1410	1591	181	1079	1841	762
Sheola	762	2143	1381	552	780	228
Moulvi Bazar	452	622	170	391	236	−155
Habigonj	402	577	175	393	453	60
Durgapur	852	978	126	608	751	143
Bhairab Bazar	432	613	181	325	292	−33
Comilla	492	813	321	369	328	−41
Chandpur	380	736	356	320	538	218
South Eastern Hill Basin						
Parshuram	557	831	274	414	577	163
Chittagong	680	1247	567	568	1306	738
Panchpukuria	645	1203	558	462	1046	584
Bandarban	612	880	268	424	690	266

4.2. THE DISCHARGE, PERIOD OF OCCURRENCE, WATER LEVEL AND DURATION REGIMES

Peak discharge values and corresponding water levels in the major rivers are the major contributors to riverine flooding in Bangladesh. However, when a peak discharge value is assessed in terms of external and internal runoff contributions, location of the gauging station must be taken into account. The stations at Hardinge Bridge (Ganges River) and Bahadurabad (Brahmaputra River) are close to the Bangladesh-India border (Figure 1). Basin areas above these two stations within Bangladesh are small and it is unlikely that these small parts of basins contribute significantly to generate peak discharges of the Ganges and Brahmaputra rivers. However, the Bhairab Bazar station on the Meghna River is located far from the border and receives significant amounts of runoff from the area within Bangladesh.

4.2.1. *Peak Discharge*

The peak discharge of the *Ganges River* in 1987 at Hardinge Bridge was the highest ever at that station. The computed peak discharge at this station was 76,000 m^3/sec, which exceeded the previously recorded highest flow of 73,200 m^3/sec in 1961. The peak flow for the Brahmaputra River at Bahadurabad in 1998 was 102,534 m^3/sec, the highest ever, slightly more than the previous record of 98,600 m^3/sec in 1988. Note that while the Ganges River drains a basin area twice the size of the Brahmaputra basin, the magnitude of peak discharge of the Brahmaputra River is greater than the Ganges River because average monsoon rainfall in the later is twice that of the former. The highest peak discharge of the *Meghna River* at Bhairab Bazar occurred in 1988 (19,800 m^3/sec), slightly higher than the previous recorded highest flow of 19,500 m^3/sec in 1974. Magnitudes of peak discharge and water levels and their return periods are tabulated in Figure 5 and Figure 6 which shows that the Brahmaputra River produces the most extreme floods followed by the Ganges River.

4.2.2. *Period of Occurrence*

One important aspect of the 1987 flood was that the peak flows of the Ganges and Brahmaputra did not coincide. For the Brahmaputra River, it was about period one as described in Section 3. While the Ganges peak occurred 15 days later (19 September) than the average peaking date September 4 (Figure 13). Peak discharge in the Ganges and Brahmaputra occurred 34 days apart and the likelihood of such occurrence is 21.5% (Table II). The pattern was similar to the floods in 1954 and 1974 (Matin, 1987). The Meghna peak occurred on 8 August, just 4 days prior to the Brahmaputra peak. This implies similar rainfall patterns in the two river basins in 1987.

In 1988, the peak discharges of the three rivers occurred about the same time (Figure 13). For the Brahmaputra River, the date of occurrence of peak discharge falls around period two (see Section 3). As indicated before, during period two, the

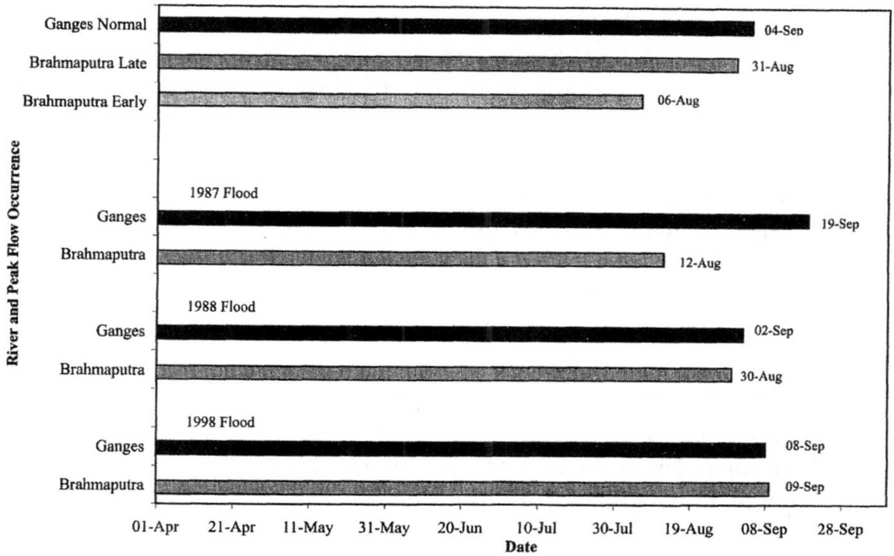

Figure 13. Date of occurrence of flood peaks.

likelihood of a simultaneous occurrence is very high. The Ganges peak occurred just two days earlier than the average peaking date (Figure 13). Peak discharge in the Ganges and Brahmaputra occurred only three days apart. The combined flow of the two rivers slowed the outflow from the Meghna River at Chandpur. The high floods on the Ganges and Brahmaputra rivers, their similar times and the backed up flood water in the Meghna basin caused a severe flood in Bangladesh.

In 1998, the times of occurrence of the peak discharges of the Ganges, Brahmaputra and Meghna rivers (Figure 13) had two important aspects. *First*, they occurred in the same month in September. *Second*, flood peaks in the Ganges and Brahmaputra rivers occurred only a day apart, an occurrence, which had not happened before in recorded history (Figure 13). The time of occurrence is unusual for the Brahmaputra, as only 15% of the flood peaks occurred in September (Table I). The Meghna flood peak also occurred on 7 September. The simultaneous occurrence of flood peaks in the three rivers caused the severe flood, which engulfed 69% of the country.

4.2.3. *Water Levels*

In 1987, all the monitoring stations in the Ganges basin reported exceedence of danger levels. This implies severe flooding in the Ganges basin. In the Brahmaputra basin, water levels were above the danger level in 7 stations out of 10. In the Meghna basin the situation was as serious as the Ganges basin. In the South Eastern Hill Basin, water levels exceeded danger levels in all stations. Overall, in 92% of the monitoring stations (31 out of 34) of the four basins, flood level exceeded danger levels (BWDB, 1987).

In 1988, the water level in all but one monitoring station on the Ganges River crossed the danger level. Although peak discharge was slightly lower than in 1987, the monitoring stations Hardinge Bridge, Goalundo, Bhagyakul, Gorai Railway Bridge and Faridpur registered higher water levels than in the previous year, perhaps caused by deposition of sediments from the high floods of 1987. Due to very high flood flows in the Brahmaputra River and in its tributaries, water levels exceeded respective danger levels in all monitoring stations as did the Meghna basin also registered water levels above danger levels. A similar situation was reported on the South Eastern Hill Basin. In total, water levels in 97% of the monitoring stations exceeded danger levels, demonstrating the severity of flood in 1988 in Bangladesh.

The flood situation of 1998 became worse on July 28 when 23 stations monitoring several rivers were reported flows above their respective danger levels (Islam and Chowdhury, 1999). Steady or little improvement was observed between the end of July and the middle of August. The flood situation deteriorated again with all the stations in the Ganges basin exceeding their danger levels and became severe on August 30, when 24 stations in all basins were above their danger levels. After a slight improvement for about a week, the situation worsened again due to the synchronisation of the flood peaks through the upstream of Brahmaputra and Ganges rivers, and became extremely severe on September 7 when 25 monitoring stations had flows above the danger level (Islam and Chowdhury, 1999).

4.2.4. *Depth of Floods*

Analysis of depth of flooding reveals some interesting features for the major rivers. The depths of flood for the Ganges and Brahmaputra rivers were almost equal in 1987. The maximum depth of flood water at Hardinge Bridge on the Ganges was 1.3 meters (calculated by subtracting ground level[3] from the water level). At Bahadurabad, the maximum depth of flood was 1.28 m, while flood depth for the Meghna was only 0.4 m at Bhairab Bazar. However, higher flood depths were observed at various upstream stations in the Meghna Basin (BWDB, 1987).

In 1988, maximum depth of flooding in the Ganges River at Hardinge Bridge was estimated to be 1.37 m. This was 7 cm higher than the flood depth of 1987 although the peak discharge was smaller. A staggering 2.22 m flood depth was observed for the Brahmaputra River at Bahadurabad. The depth of flooding demonstrates the severity of flooding in 1988 in the Brahmaputra basin. Depth of flooding for the Meghna at Bhairab Bazar was approximately 1.16 m, which was 0.8 m higher than in 1987, caused by the high peak discharge of the Meghna River. The Meghna River at Chandpur registered a maximum 1.86 m flood depth.

Although in 1998, the magnitude of flood peak of the Ganges River at Hardinge Bridge was smaller than that of 1988, the recorded water level was higher Figure 5.

[3] Ground level of the Ganges and Brahmaputra at Hardinge Bridge and Bahadurabad is 13.5 + PWD and 18.4 m + PWD, respectively.

This was perhaps caused by aggradation/change in the channel cross-section of the river. Therefore, depth of flooding increased to 1.69 m compared to 1988. On the other hand, despite the highest ever discharge in 1998, flood depth in Brahmaputra at Bahadurabad was 1.97 m, which was 25 cm smaller than in 1988. Degradation in the river caused smaller flood depth, otherwise the flood situation would have been much worse. These phenomena demonstrate that morphological conditions also control the flooding process in the large rivers in Bangladesh. Flood depth in the Meghna River at Bhairab Bazar was 1.08 m, against a lower peak discharge than 1988.

4.2.5. *Flood Duration*

The duration of floods is important with regard to damage and recovery. In 1987, the duration of water levels above the danger levels in all stations of the Ganges River exceeded previous records mainly due to the highest ever peak discharge. One spectacular aspect is that water levels at Hardinge Bridge, Goalundo and Bhagyakul stayed above respective danger levels for equal periods of time (Table VIII). This was probably caused by a constant rate of drainage downstream. Figure 3 illustrates the relationship between depth of water above danger level and recession time. As the water levels were significantly high, it took a long time for them to recede to below the danger level. For example, the danger level at Chilmari for the Brahmaputra stayed for 24 days. This was caused by the backwater effect at the confluence of the Brahmaputra and the Teesta rivers. A similar backwater effect at the confluence of the Ganges and Brahmaputra rivers caused the water level to remain above danger level at Serajgonj for 41 days. The Kushiyara (upstream of the Meghna) at Sheola and the Meghna at Bhairab Bazar remained above danger levels for almost an equal period of time probably caused by a constant rate of drainage.

In 1988, the duration of floods above danger levels at various monitoring stations in the Ganges basin was lower than that of 1987 due to a smaller peak discharge (Table VIII). Once the flood peak was reached, recession occurred promptly. However, in the downstream stations, a slow rate of recession was observed. For example, while the flood level was only seven days above the danger level at Hardinge Bridge in September, it stayed 17 and 18 days at Goalundo and Bhagyakul, respectively. The duration of floods was almost equal for the Brahmaputra at Bahadurabad for the months of August and September (9 and 11 days, respectively). In total at Bahadurabad, the Brahmaputra remained above the danger level for 27 days (Table VIII and Figure 14). The duration of flooding for the Meghna River was spectacular. At Bhairab Bazar it stayed above the danger level for 68 days, which was significantly higher than the other two rivers (Table VIII and Figure 14). The floods ended in the Ganges and Brahmaputra rivers by the middle of September, while flooding in the Meghna basin continued until the end of September. This was caused by backwater effect and build up of water in the downstream reaches of the combined rivers. A spring tide in conjunction with a solar eclipse on 11 September

Table VIII. Duration (in days) of floods above the danger levels in 1987, 1988 and 1998

River	Station	1987	1988	1998
Ganges	Rajshahi	29	24	28
	Hardinge Bridge	55	23	27
	Goalundo	54	41	68
	Bhagyakul	56	47	72
Gorai	Gorai Railway Bridge	43	25	25
Brahmaputra	Chilmari	24	15	22
	Bahadurabad	13	27	68
	Serajgonj	41	44	48
	Aricha	–	31	68
Teesta	Kaunia	42	38	–
Meghna	Bhairab Bazar	30	68	68
Surma	Kanairghat	–	75	73
Kushiyara	Sheola	27	80	37

considerably delayed the recession of flood water from the southern part of the country (Hofer, 1998).

The most significant feature of the 1998 flood was the long duration of floods above danger levels in all the major rivers. In terms of duration, the floods of 1998 surpassed all past records (Ahmed and Mirza, 2000). (Table VIII and Figure 14). From the middle of August to the middle of September, flood levels were consistently above danger level on the Ganges at Hardinge Bridge. The case of the Brahmaputra and Meghna was the worst ever. For most of July and from the middle of August to the middle of September, a total of 68 days, flood levels of the Brahmaputra were above danger levels at Bahadurabad (Table VIII and Figure 14), while for the Meghna, it was 68 days (from the middle to July to late September). The duration of the flood in one upstream station of the Meghna River was a week longer than Bhairab Bazar.

What caused such a long duration of floods in the three rivers? The flood peaks of the Ganges and Brahmaputra rivers occurred only 24 hours apart. The combined flow had difficulty passing through the narrow cross-section of the Padma River downstream of Goalundo, which often causes a backwater effect that can be observed about 30 kilometers upstream of the confluence of the Ganges and Brahmaputra rivers (RPT-NEDECO-BCL, 1993). Lunar position (full moon) and its strong influence often cause high tides and a subsequent backwater effect. The spring tide during the monsoon of 1998 was much higher than usual, causing tremendous backwater effect and eventually inhibiting drainage. Sea level in the northern Bay of Bengal had played a role in flooding. In 1998, sea-level height

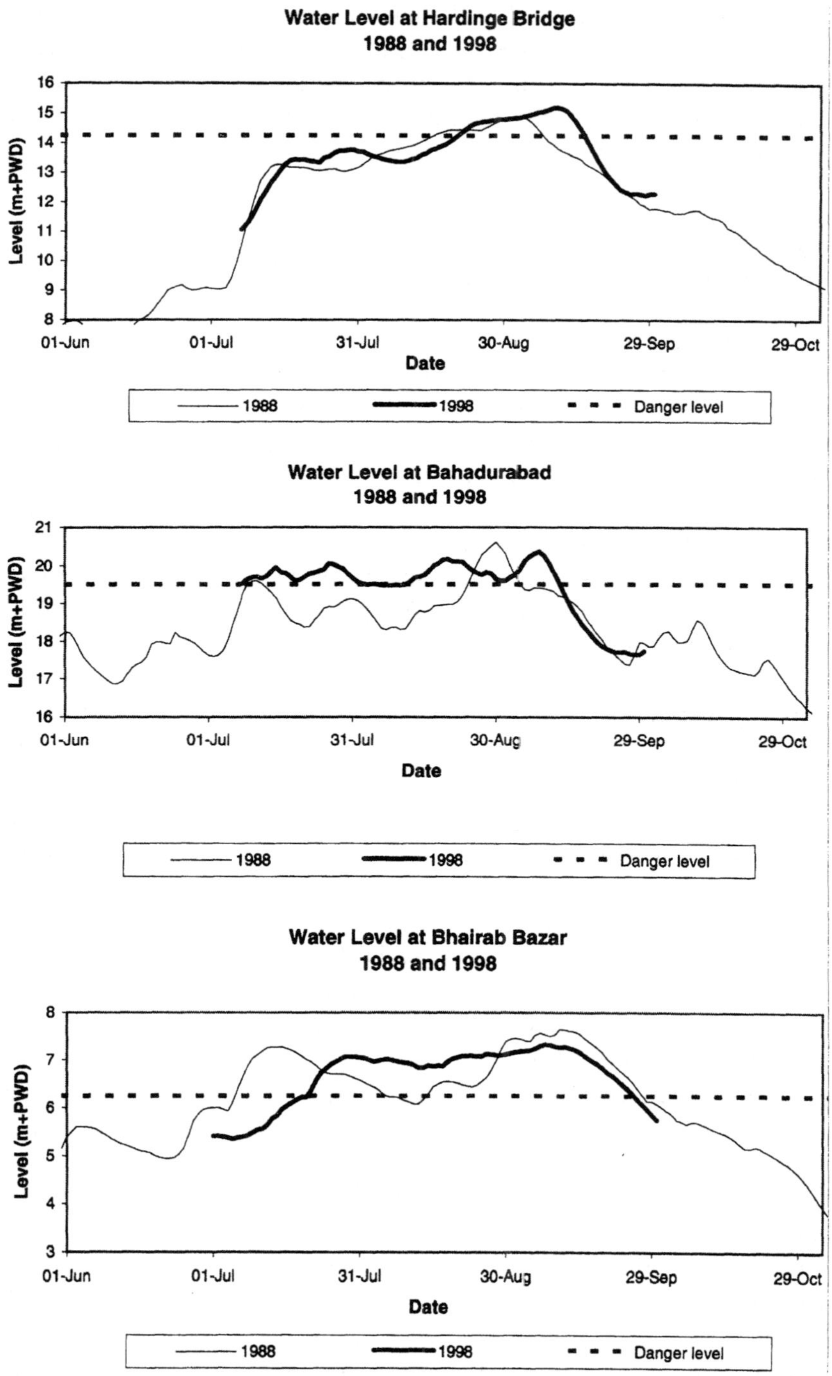

Figure 14. Water level and flood duration in the GBM rivers in Bangladesh for 1988 and 1998.

in the northern Bay of Bengal was 0.33 m higher than in 1997 (Peter J. Webster, University of Colorado, personal communications). The compounding effect was 'stagnation' of water between the Ganges and Meghna confluence points. Consequently, a vast area in the central and north-eastern parts of Bangladesh was submerged for an unusually long time, resulting in the deluge of the century (Ahmed and Mirza, 2000).

5. The Three Extreme Floods: Conclusion

Basinwide monsoon rainfall patterns control flood peaks of the Ganges, Brahmaputra and Meghna rivers. The progression of monsoon air mass is from the southeast part of the GBM basin to the northwest. Therefore, the Brahmaputra and Meghna basins experience their first flood peaks in July, while the Ganges basin usually receives intense rainfall in August- September that leads to its flood peak. However, the Brahmaputra and Meghna rivers may also experience high rainfall at the same time of the Ganges River in the upstream and downstream areas.

Along with local rainfall, rainfall in the basin areas adjacent Bangladesh generally control flood peaks and flooding process. Local rainfall generally aggravates the flooding situation in the downstream areas subject to timing of rainfall occurrences.

Rainfall in a drier basin generally has higher variability than in a wetter basin. Among the three river basins, the variability of rainfall in the Ganges basin is higher than in the Brahmaputra and Meghna basins. This variability causes high uncertainty in the peak discharge of the Ganges basin than the other two basins. This has implications for flood damage and mitigation planning in the Ganges basin.

Due to rainfall patterns in the basins and the time of occurrence of flood peaks, synchronisation of flood peaks may happen in Bangladesh although not very often. However, the flood peaks of the Ganges and Brahmaputra rivers coincided twice in the case of the three extreme floods analysed in this paper, and led to extreme flooding in Bangladesh in 1988 and 1998.

The narrow cross-section of the Padma River (the combined the Ganges and Brahmaputra rivers) cannot quickly drain the combined flood volume, which may be as high as 180,000 m3/sec. Observations show that the drainage of flood flow is further slowed due to higher sea-level in the northern Bay of Bengal, and the high tide caused by lunar activity. This is particularly important for flooding in the central part and Meghna basin in north-eastern Bangladesh.

High flood levels in the main rivers indicate a longer flooding period in Bangladesh. However, many other factors such as constant supply of flood flow, coincidence of flood peaks, tidal activity, sea-level, etc. control flood duration. The synergic effects of these factors on flood duration are difficult to estimate, and call for a detailed analysis.

Acknowledgements

I appreciate comments from David Etkin and Marie Sanderson of the Adaptation and Impacts Research Group (AIRG) and two anonymous reviewers on an earlier draft of the paper. Thanks to Farzana Abdulhusein of the AIRG for redrawing most of the figures.

References

Ahmed, A. U. and Mirza, M. M. Q.: 2000, Review of causes and dimensions of floods with particular reference to flood'98: National perspectives. In: Q. K. Ahmad, A. K. A. Chowdhury, S. H. Imam and M. Sarker (eds.), *Perspectives on Flood 1998*, University Press Ltd., Dhaka.

Ahmed, R.: 1999, Monsoon prediction in Bangladesh using early pre-monsoon interior-to-coast temperature gradient index (TGI) and southern oscillation index (SOI). Paper presented at the Conference on Numerical Weather Prediction (NWP), Jadavpur University, Kolkata, India, December 30, 1999.

BWDB (Bangladesh Water Development Board): 1987, *Flood in Bangladesh – 1987*, BWDB, Dhaka.

Boyce, J. K.: 1990, Birth of a megaproject: Political economy of flood control in Bangladesh, *Environmental Management* **14**(4), 419–428.

BWDB (Bangladesh Water Development Board): 1995, *Discharge Data for Various Rivers in Bangladesh*, BWDB, Dhaka.

Brammer, H.: 1990, Floods in Bangladesh I. Geographical background of the 1987 and 1988 floods, *The Geographical Journal* **156**(1), 12–22.

Chowdhury, A. M.: 2000a, Flood, 98: Oceanic perspective, In: *Perspectives on Flood 1998*, In: Q. K. Ahmad, A. K. A. Chowdhury, S. H. Imam and M. Sarker (eds.), University Press Ltd., Dhaka.

Chowdhury, M. R.: 2000b, An assessment of flood forecasting in Bangladesh: The experience of the 1998 flood, *Natural Hazards* **22**, 139–163.

CSE (Centre for Science and Environment): 1992, *Floods, Flood Plains and Environmental Myths*, CSE, New Delhi.

Dhar, O. N. and Kamte, P. P.: 1973, Probable maximum precipitation over the Brahmaputra basin in Assam, *Irrigation and Power Journal* **30**(3), 237–241.

Dhar O. N. and Nandargi, S.: 2001, A comparative flood frequency study of Ganga and Brahmaputra river systems of north India-a brief appraisal, *Water Policy* **3**, 101–107.

Dhar, O. N., Soman, M. K., and Mulye, S. S.: 1984, Rainfall over the southern slopes of the Himalayas and the adjoining plains during 'breaks' in the monsoon, *Journal of Climatology* **4**, 671–676.

FEC (French Engineering Consortium): 1989, *Prefeasibility Study for Flood Control in Bangladesh – Vol. 2: Present Conditions*, FEC, Paris.

FFWC (Flood Forecasting and Warning Centre): 1998, Danger Levels and Flood Duration (http://www.ffwc.net/html/danger_rh_wl.htm).

FAP (Flood Action Plan) 25: 1992, *Flood Hydrology Study*, Flood Plan Corrdination Organisation (FPCO), Dhaka.

GOI (Government of India) and Government of Bangladesh (GOB): 1990, *Indo-Bangladesh Task Force on Flood Management Report*, GOI and GOB, New Delhi and Dhaka.

Hofer, T.: 1998, Floods in Bangladesh: A highland-lowland interaction? Institute of Geography, University of Berne, Switzerland.

IITM (Indian Institute of Tropical Meteorology): 1989, *Probable Maximum Precipitation Atlas*, IITM, Pune, India.

IMD (Indian Meteorological Department): 2001, *Subdivisionwide Monsoon Precipitation*, IMD, New Delhi.

Islam, S. M. and Chowdhury, M. R.: 1999, 1998 flood in Bangladesh-Obscure and eventualities of nature, *Proceedings of IDNDR Symposium on Mitigation of Water Related Disasters* held at Nagoya, Japan, February 16–18.

Islam, S. R. and Dhar, S. C.: 2000, *Bangladesh Floods of 1998: Role of Flood Forecasting & Warning Centre*, BWDB, Dhaka.

Krishnamurti, T. N., Bedi, H. S., and Subramaniam, M.: 1987, *On the Summer Monsoon of 1987*, Tallahassee, Florida.

Kulshrestha, S. M.: 1997, *Drought Management in India and Potential Contribution of Climate Prediction*, Center for Ocean-Land-Atmosphere Studies and Center for the Application of Research on the Environment, Maryland, USA.

Matin, M. A.: 1987, Some aspects of 1987 flood. Presented in Seminar *Floods in Bangladesh*, held in the Institution of Engineers, Bangladesh (IEB), Dhaka, Bangladesh.

Matin, M. A. and Husain, M. A.: 1988, Hydrological aspect of 1988 flood, Presented in Seminar *Floods in Bangladesh*, held in the Institution of Engineers, Bangladesh (IEB), Dhaka, Bangladesh.

Matsumoto, J.: 1988, Synoptic features of heavy monsoon rainfall in 1987 related to the severe flood in Bangladesh, *Bulletin of the Department of Geography* **20**, 43–56.

Mirza, M. M. Q.: 1991, Flood action plan of Bangladesh – The Embankment Issue, *Water Nepal* **2**(2/3), 25–28.

Mirza, M. M. Q. and Paul, S.: 1992, *Pakritik Durjog O Bangladesher Paribesh* (in Bengali), Centre for Environmental Studies and Research (CESR), Dhaka.

Mirza, M. M. Q.: 1997, *Modelling the Effects of Climate Change on Flooding in Bangladesh*, Unpublished PhD. Thesis, International Global Change Institute (IGCI), University of Waikato, Hamilton, New Zealand.

Miyakoda, K., Kinter, J., and Yang, S.: 2000, *Analysis of the Connection from the South Asian Monsoon to ENSO by Using Precipitation and Circulation Indices*, COLA Technical Report 90. Center for Ocean-Land-Atmosphere Studies, Calverton, USA.

Parthasarathy, B. and Pant, G.B.: 1985, Seasonal relationships between Indian summer monsoon rainfall and the southern oscillation, *J. Climatol.* **5**, 369–378.

Popeleswski, C. F.: 1988, The global climate for June-August 1988: A swing to the positive phase of the southern oscillation, drought in the United States, and abundant rain in monsoon areas, *J. Climate* **1**(11), 1153–1174.

Rao, Y. P.: 1976, *Southwest Monsoon*, New Delhi, India.

Rao, Y. P.: 1981, The climate of the Indian subcontinent, In: K. Takahashi and H. Arakawa (eds.), *Climate of Southern and Western Asia. World Survey of Climatology*, Vol. 9, Elsevier, Amsterdam.

Relief Web: 2001, India-Floods and Landslides OCHA Situation: Report No. 3 (http://wwwnotes, reliefweb.int/websites/rwdomino.nsf/86d6cf0398b9c2fc125650a00512b.

Rogers, P., Lydon, P., and Seckler, D.: 1989, *Eastern Waters Study: Strategies to Manage Flood and Drought in the Ganges-Brahmaputra Basin*, ISPAN, Virginia, USA.

RPT-NEDECO-BCL: 1993, *Jamuna Bridge Study, Phase II Study, Feasibility Report, Annex B: River Morphology*, Jamuna Multipurpose Bridge Authority, Dhaka.

Schneider, B.: 1996, *Drei Fallstudien von Niederschlagsereignissen der Uberschwemmungsjahre 1987/88 in Bangladesh*, Eine Analyse von Wolkenstrukturen auf NOAA-Satellitenbilder und Klimatologisschen Daten, Unpublished Masters Thesis, Department of Geography, University of Berne.

UNDP (United Nations Development Programme): 1989, *Bangladesh Flood Policy Study-Final Report*, UNDP, Dhaka.

Natural Hazards **28:** 65–84, 2003.

Geomorphic Effects of Monsoon Floods on Indian Rivers

VISHWAS S. KALE
Department of Geography, University of Pune, Pune 411 007, India
E-mail: vskale@unipune.ernet.in

(Received: 22 August 2001; accepted: 4 March 2002)

Abstract. The southwest summer monsoon contributes the bulk of India's rainfall. Consequently, almost all the geomorphic work by the rivers is carried out during the monsoon season in general and the monsoon floods in particular. Indian rivers are characterized by high average flood discharges and large temporal variability. There is also significant spatial variation in the magnitude, frequency and power of floods, on account of regional variations in monsoon rainfall, basin characteristics and channel geometry. As a result, the channel responses and the geomorphic effects also vary spatially. This paper describes the hydrological and geomorphological aspects, as well as the geomorphic effects of monsoon floods in the Indian rivers. The geomorphic effects of floods are most impressive only in certain areas – the Himalaya, the Thar Desert, and the Indus-Ganga-Brahmaputra Plains. There are numerous instances of flood-induced changes in the channel dimension, position and pattern in these areas. In the Ganga-Brahmaputra Plains, the annual floods appear to be geomorphologically more effective than the occasional large floods. In comparison, the rivers of the Indian Peninsula are, by and large, stable and the geomorphic effects of floods are modest. Only large-magnitude floods that occur at an interval of several years to decades are competent to modify the channel morphology in a significant way. A synthesis of the various case studies available from the Indian region indicates that often the absolute magnitude of a flood is not as important with respect to the geomorphic effects as the flow stress and competence.

Key words: Monsoon, floods, geomorphic effects, channel changes, India

1. Introduction

The southwest summer monsoon contributes more than eighty percent of the annual rainfall over a major part of the Indian region. Only in the southeastern part of the Indian Peninsula, and in the northwest Himalaya the proportion drops to less than 60% (Figure 1). Therefore, over most of the Indian region the rivers carry out almost all the geomorphic work of erosion, transportation and deposition during the summer monsoon season. Due to the seasonality of flows, the average monsoon flows tend to be at least 1–2 orders of magnitude more that in the non-monsoon season. Consequently, with the arrival of monsoon rains there is sudden revival of the fluvial activity over much of the Indian region. Rivers do not, however, remain at a high stage throughout the monsoon season. It is only after a spell of heavy rains, which may last for a period of several hours to several days, that large

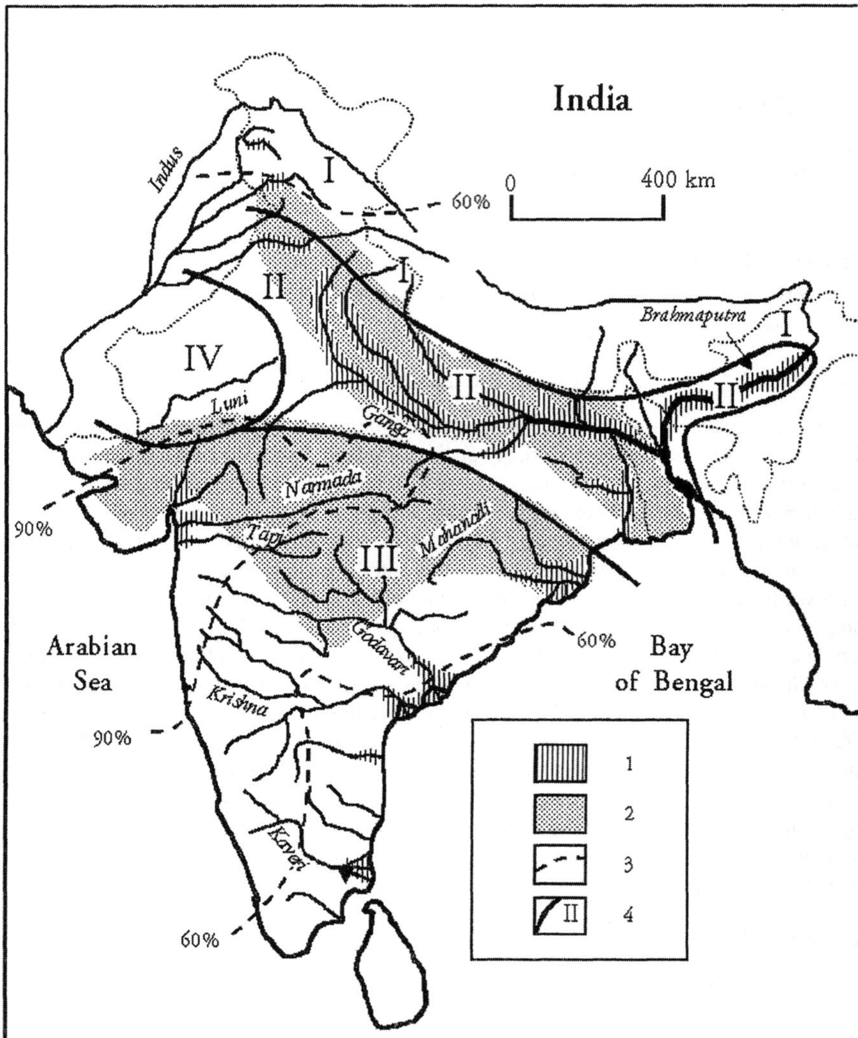

Figure 1. Map of India showing the major rivers and the flood-prone areas. 1 = flood-prone areas; 2 = major rainstorm zones after Dhar and Nandargi (1993); 3 = percentage of annual rainfall contributed by southwest monsoon; 4 = major hydro-geomorphic zones mentioned in the text (I–IV).

runoff is generated in the catchments, and the rivers experience floods. As a result of these rainfall characteristics, the rivers of the Indian region display a channel morphology that is largely adjusted to (Gupta, 1995; Kale, 1998)

1. Wide range of discharges – non-monsoon low flows to monsoon high flows.
2. Occasional large-magnitude floods that occur at an interval of several years to decades.

3. Pulses of sediment flux imposed from upstream and by the tributaries, and
4. Channel perimeter and floodplain lithology/sedimentology.

Therefore, in spite of the large spatial variation in the size and periodicity of large floods, all rivers, big or small, display channel forms and features related both to the common monsoon flows and the infrequent large-magnitude floods (Gupta, 1988).

This paper describes the hydrological and geomorphological aspects of flooding in the Indian rivers. The geomorphic effects of floods in different hydro-geomorphic regions of India are also discussed.

2. Flood Hydrology

Because the precipitation over a large part of India is strongly concentrated in the summer monsoon season during June and September/October, and the tropical storm season from May to October, floods occur during any part of the monsoon season. However, they are more frequent in August and September. There is marked spatial variation in the magnitude, frequency and duration of floods on account of the differences in the catchment size, basin relief and basin location with reference to the monsoon winds and cyclone tracks (Gupta, 1988; Kale, 1998). River basins, such as Narmada, Tapi, Mahanadi and lower Godavari, located in the path of tropical cyclones experience higher unit discharges than other rivers (Kale *et al.*, 1997).

Large synoptic systems ranging in force from tropical lows to cyclones are the main cause of unusually large floods on the Indian rivers (Ramaswamy, 1987). The flood-generating rainstorms are usually of 3-day duration and are confined to two major zones (Dhar and Nandargi, 1993). Whilst the Ganga Basin and the Punjab Plains fall in the first zone, central India and the northern half of the Indian Peninsula constitute the second zone. Further, some of the highest 24-h rainfalls are observed in the hilly regions of northeast India and the foothills of the Himalaya (Rakhecha and Pisharoty, 1996). Therefore, the Ganga and Brahmaputra Basins have the highest flood potential in the Indian subcontinent, followed by the rivers of central India, such as Mahanadi, Narmada, Tapi, and lower Godavari (Kale, 1998). Consequently, all rivers of these regions display flood-related forms and features.

Whilst heavy rainfall is usually the main cause of large floods on the Indian rivers, sometimes failure of natural and man-made dams are also known to generate large floods. Such floods are often associated with some of the most extreme discharges and flood power (Shroder, 1998; Kale, 1998). For example, estimates based on hydraulic modeling for the catastrophic Chandora dam-failure flood in upper Tapi River in July 1991 indicate that the discharge per unit area was more than 12 $m^3\ s^{-1}\ km^{-2}$ (Kale *et al.*, 1994). In comparison, the unit discharges of natural extreme floods on large Peninsular rivers range between 0.13 and 0.65 $m^3\ s^{-1}\ km^{-2}$ (Kale *et al.*, 1997; Hire, 2000).

The majority of Indian rivers are characterized by high average discharges, variability and positive skewness (Sakthivadivel and Raghupathy, 1978; Garde and Kothyari, 1990). For medium and large-sized basins (10^4-10^5 km^2) the 2.33-y flood is between 3,000 and 55,000 m^3 s^{-1}. The largest discharges ($Q_{\max}$) in the modern record on large Indian rivers range between 6,000 and 90,000 m^3 s^{-1}. These are about 2 to 4 times greater than the 2.33-y flood (Q_m). Whilst single-peak flood events during the monsoon season are a common feature, sometimes successive flood-generating meteorological conditions may give rise to multiple peak events. Such situations occur more frequently in the Ganga and Brahmaputra Basins as well as in central India.

Variability is an integral part of the fluvial regime of the Indian rivers, with sometimes enormous interannual (Figure 2) and intra-monsoonal variations in the flood magnitudes (Sakthivadivel and Raghupathy, 1978; Garde and Kothyari, 1990; Gupta, 1995; Kale and Gupta, 2001). On account of large interannual variability in the amount, intensity and duration of monsoon rainfall, there is also considerable variability in the duration, volume and magnitude of monsoon flows from year to year and from one decade to the other (Kale, 1999). Figure 2 shows the long-term interannual variability in the annual peak discharge/stage for some major rivers in India. The figure shows that while the variability in the peak discharge/stage was relatively lower between 1900s and 1940s, the variability has increased significantly since 1950s (Kale, 1999). The pattern depicted by Kaveri River (Figure 2F) is different because the river basin lies in a different rainfall regime zone, where the bulk of annual rainfall is received during the winter monsoon season.

One widely used measure of the flood variability is the flash flood magnitude index (FFMI), which is the standard deviation of the logarithms of the annual maximum discharge (Baker, 1977). Figure 3 gives a plot of the FFMI values against $Q_{\max}/Q_m$ ratios. The plot reveals that rivers either in the high relief setting (such as Teesta, upper Yamuna and upper Ganga) and/or in the major rainstorm zones (e.g., Tapi, Godavari, Narmada, Damodar) display higher $Q_{\max}/Q_m$ ratio and FFMI values. On the other hand, lower Brahmaputra, lower Ganga and Krishna Rivers draining the plains are characterized by lower interannual variability in flood discharges. Figure 3 also shows that the FFMI values for most large Indian rivers are lower than the world average, implying that year-to-year differences in peak flood magnitudes are not large. In other words this suggests that large flows are relatively frequent in Indian rivers. In addition to the inter-annual variability, the pattern of discharge in any one year is also important from the standpoint of geomorphic effectiveness. Analyses of recent gage data by Dhar and Nandargi (1998) indicate that the number of floods on the Brahmaputra River during a monsoon season vary from 5 to 19. This is to say that on average the frequency of floods varies between once in 4–5 weeks and once in 10 days during the monsoon season. The corresponding values for Ganga and Kosi Rivers are 4–10 and 5–14 respectively (Dhar and Nandargi, 1998). These data imply that floods dominate the forms and process of the rivers of the Indus-Ganga-Brahmaputra (IGB) Plains.

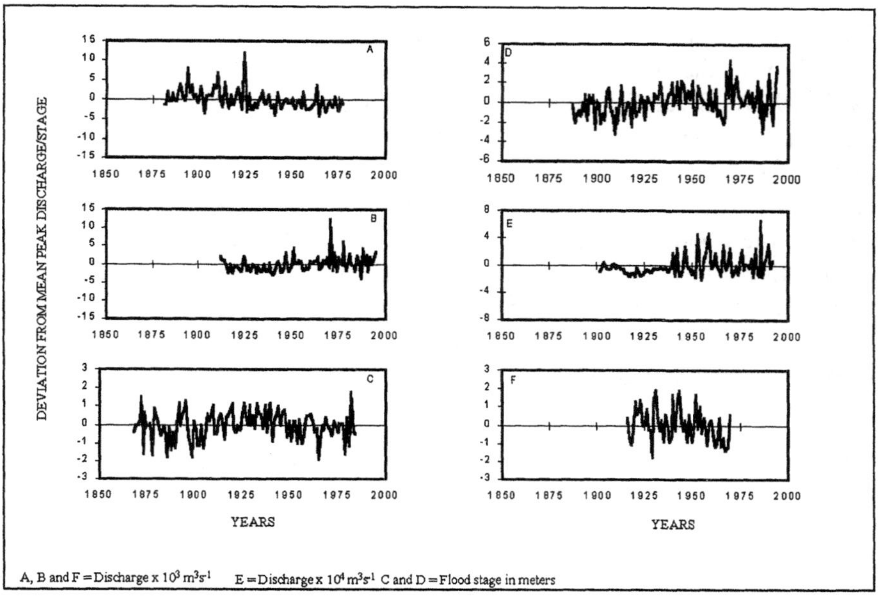

Figure 2. Plot of deviation of annual peak discharge/stage from mean for selected rivers of India. A = Ganga (upper) at Raiwala; B = Satluj at Bhakra; C = Mahanadi at Naraj; D = Narmada at Broach; E = Godavari at Dowleshwarm; F = Kaveri at Grand Anicut.

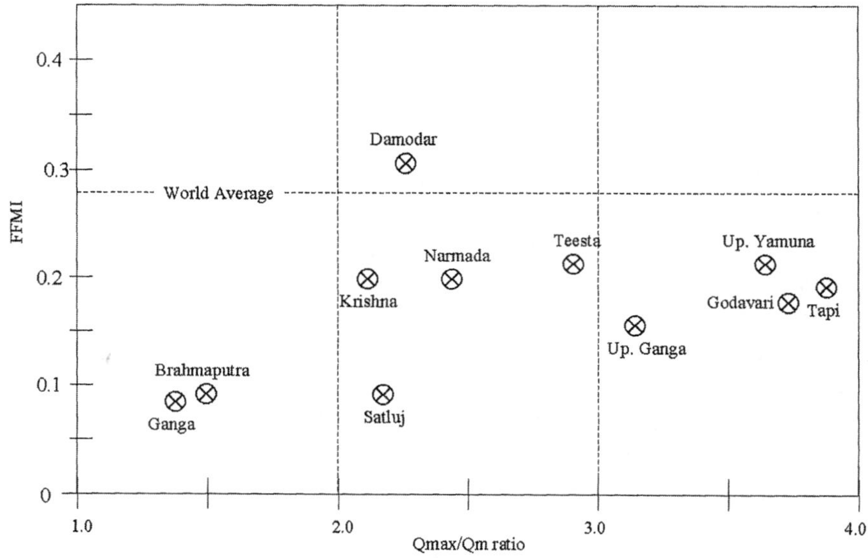

Figure 3. Plot of the ratio between largest discharge on record (Q_{max}) and mean annual peak discharge (Q_m) against the flash flood magnitude index (FFMI) for some selected Indian rivers.

3. Flood Geomorphology

The monsoon-dominated rivers, conform more closely to environments with high variability in discharge, where large floods are the formative discharges (Gupta, 1995, 1998). Given the high magnitude and frequency of floods in the Indian region it is reasonable to conclude that most monsoon floods have the ability to alter floodplain and channel morphology in a significant way. However, several studies have shown that the flood magnitude is not as important with regard to geomorphic work as the flood power (Baker and Costa, 1987; Kochel, 1988; Richards, 1999). Floods likely to result in significant geomorphic changes are those that produce high shear stress to entrain the bed sediments and to exceed the resistance provided by bed and banks (Baker and Costa, 1987). Unit stream power or boundary shear stress, therefore, are good measures of the geomorphic effectiveness of a flood.

3.1. FLOOD POWER AND SEDIMENT TRANSPORT COMPETENCE

The flood power is defined in terms of channel boundary shear stress and power per unit area of bed (Baker and Costa, 1987):

$$\omega = \gamma QS/w = \tau V \tag{1}$$

$$\tau = \gamma RS \tag{2}$$

where, ω is unit stream power expressed in watts per square meter (W m^{-2}), γ is the specific weight of clear water (9800 N m^{-2}), Q is discharge in m^3 s^{-1}, S is the slope, w is the water surface width in m, τ is the boundary shear stress expressed in Newtons per square meter (N m^{-2}), R is the hydraulic radius or mean depth of water in m, and V is velocity in m s^{-1}.

Calculations indicate that the estimated average unit stream power at a cross-section (alluvial) varies between 322 W m^{-2} on Godavari and less than 10 W m^{-2} on the Brahmaputra (Kale, 1998). In case of the Godavari River this is mainly due to low width-depth ratio and relatively higher channel gradient. In comparison the estimated power per unit area of bed for the largest floods on the alluvial Amazon and Mississippi Rivers is about 12 W m^{-2} (Baker and Costa, 1987) (Figure 4). In the bedrock reaches, however, the unit stream power values are higher by one or two orders of magnitude than those that occur in the alluvial channels (Kale *et al.*, 1994; Rajaguru *et al.*, 1995; Baker and Kale, 1998).

During large monsoon floods on alluvial rivers, such as Ganga and Brahmaputra, the values of power per unit area are small due to drastic width increase by overbank flows. The width, which is inversely related with the power per unit area, also increases by bank erosion and channel widening. Coleman (1969) noted that during passage of floods, the channel area increases by about 300% in the lower Brahmaputra River. Similar, changes in the channel width and area have been observed for other rivers of the IGB Plains.

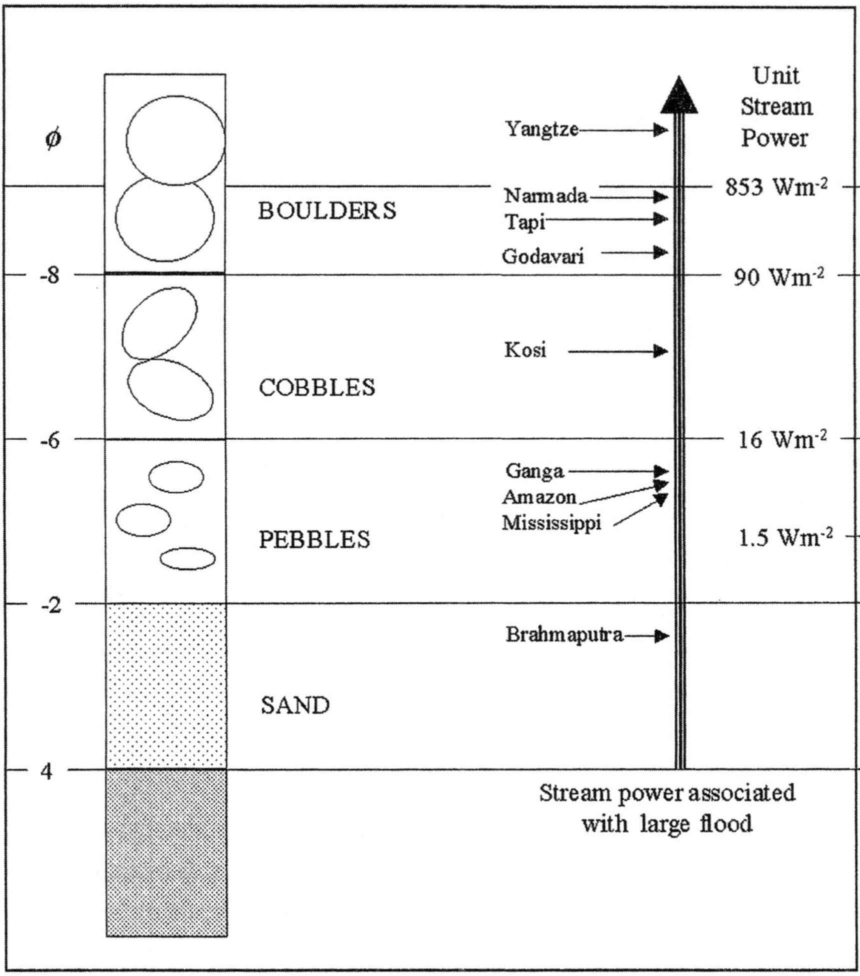

Figure 4. Unit stream power associated with large floods on some Indian and world rivers. The threshold of movement of coarse sediments in terms of unit stream power is also given for reference. The thresholds were estimated by using empirical relationship ($\omega = 0.079d^{1.27}$, where d is intermediate diameter of particle in mm) for coarse sediment ($10 < d < 1500$ mm) transport developed by Williams (1983). Data for Yangtze (1870 flood), Amazon and Mississippi Rivers after Baker and Costa (1987).

According to Komar (1988) floods generating shear stress of 10 N m^{-2} can entrain and transport sediments of about 1 cm and can carry sediments finer than 1 mm in suspension. Estimation of boundary shear stress for large floods by using Equation (2) indicates that the values for Ganga, Yamuna, and Kosi are between 11 and 25 N m^{-2} and for the Narmada and Godavari Rivers the shear stress values are above 50 N m^{-2} (Kale and Gupta, 2001), implying that most large floods on the alluvial rivers are capable of moving sand and pebbles, and sometimes even cobbles (Figure 4). However, the Brahmaputra and Dihing Rivers in northeast India have low boundary shear stress (<10 N m^{-2}) (Kale and Gupta, 2001), and thus are incapable of entraining and transporting sediment coarser than 1 cm as per Komar's (1980) estimates. This is mainly on account of very low channel gradient, high width-depth ratio and low hydraulic radius.

3.2. FLOODS AND SEDIMENT TRANSPORT CAPACITY

Several workers have noted that floods can carry a very high proportion of the annual load. Study of the Brahmaputra River shows that high flows that recur at least once every two years carry more than 95% of the annual load and flows with a recurrence interval of 10 years carry about 3% of the annual load (Goswami, 1988). More or less, similar results have been obtained for other rivers (Abbas and Subramanian, 1984; Sarma, 1986; Ramesh and Subramanian, 1988; Chakrapani and Subramanian, 1990).

Exceptional floods show a marked increase in the flow capacity. This is reflected in the high suspended sediment concentration and sediment load. Sharma and Vangani (1992) have reported high sediment concentration (0.9–40.2 g/l) during the 1979 flash flood on the Luni River in Rajasthan. The Godavari carried about 12% of the entire annual sediment load in one day during the June 1978 flood (Bikshamaiah and Subramanian, 1980). In the Balasan River in Sikkim Himalaya, the flood of September 1991 transported about 64% of the annual suspended load in two days (Starkel *et al.*, 1998). A study of the sediment transport in the Burhi Dihing River, a tributary of the Brahmaputra River, by Sarma (1986) shows that the heaviest loads of sediment are related to sudden flash floods and in some years it takes only 25 days to transport 50 percent of the total sediment. Hire's (2000) analysis of the suspended sediment load data of the Tapi River reveals that large floods carry about 20–70% of the total annual sediment load. Starkel (1972) has referred to the movement of large boulders measuring 6–12 m in diameter during the 1968 catastrophic flood in Darjeeling Himalaya. Field experiments of bedload movement in the Narmada River during the flood season by Thatte *et al.* (1986) indicate that pebble- and cobble-size sediments are moved by several tens of meters by floods.

In addition to these, reports of phenomenal movement of sediments by the Himalayan rivers during landslide dam-failure floods (Shroder, 1998; Sah and Mazari 1998; Paul *et al.* 2000) imply that floods are important geological process in monsoon-dominated areas.

4. Geomorphic Impacts of Floods

Floods occur with an unfailing regularity on some rivers of the Indian region. But many floods have little lasting impact on the landscape in general and the channel morphology in particular. There are, however, many instances of significant modifications of the channel and floodplain morphology by large floods. On some rivers the effects of large-magnitude floods persist for a long period of time. Synthesis of the several case studies available for the Indian rivers indicates that the following types of flood effects are commonly observed (Gupta, 1988) – (a) widening of channel, (b) Erosion of bars and the formation of chutes, (c) scouring of floodplain, (d) increase in competence, (e) deposition of coarse gravel within the channel and floodplain, and (f) formation of channel-in-channel topography.

Apart from the hydraulics of flows, the geomorphic effects of floods also vary with the channel geometry and perimeter lithology (Kale, 1990; Richards, 1999). Accordingly, the response of bedrock channels and the alluvial channels differs remarkably for comparable flood magnitudes. Rajaguru *et al.* (1995) and Gupta *et al.* (1999) have noted striking contrasts in the forms, features and responses between alluvial and non-alluvial reaches on the same river. Table I summarizes the geomorphic effects of floods in alluvial and bedrock channels observed in the Indian rivers.

4.1. FLOOD EFFECTS IN VARIOUS HYDRO-GEOMORPHIC REGIONS OF INDIA

Available data indicate that in terms of flood power and geomorphic effects there are noteworthy variations in the Indian region. This suggests that the characteristics of the monsoon rainfall alone cannot account for the response of rivers to floods. Other factors, such as basin geology, basin morphology and the channel geometry are also important controls on the geomorphic response. In certain parts of the Indian region, for example, the stream channel response is catastrophic due to high channel gradient, large sediment flux and easily erodible banks. In other areas the flood channel response is modest due to high resistance of the channel or low channel gradient (Kale, 1998).

For India, four hydro-geomorphic regions can be broadly demarcated that show some similarity in the geomorphic response to high flows and floods. The rivers have produced distinct channel morphology that is primarily adjusted to the basin runoff, sediment load and vegetation cover. In the following section the flood characteristics and the geomorphic effects in these hydro-geomorphic regions are described. The zones are shown in Figure 1.

4.1.1. *The Himalayan Region (Area I in Figure 1)*

Although certain parts of the Himalaya (particularly the western Himalaya) get some precipitation during the winter season also, the bulk of precipitation is received in the summer monsoon months. A high proportion of rains over the Himalaya is contributed by monsoon disturbances that originate over the Bay of Bengal or adjoining coastal areas and move in the northerly or northwesterly direction. Heavy rainfall in the Himalayan region is also associated with the 'monsoon break' situation, when the monsoon trough is located over the foothills of the Him-

Table I. Geomorphic effects of floods on Indian rivers

Geomorphic effects	Some important studies
Alluvial Channels	
• Channel deepening	Mujumdar *et al.* (1970)
• Widening of channel	Starkel (1972)
• Shifting of channels due to avulsion	Dhir *et al.* (1982)
• Erosion of bars and formation of chutes	Sarma and Basumallick (1984)
• Deposition of fines on the banks	Goswami (1985)
• Scouring of flood plains	Gupta and Dutt (1989)
• Deposition of large amount of coarse	Kale (1990)
gravel in channel	Sharma and Vangani (1992)
• Building of bars	Sinha (1996)
• Changes in channel pattern	Sah and Mazari (1998)
• Decrease in channel length due to neck cutoff	Deodhar and Kale (1999)
• Formation of a channel-in-channel	Goswami *et al.* (1999)
physiography.	Gupta *et al.* (1999)
• Increase in capacity and competence	Hire (2000)
• Formation of terrace-like features	Paul *et al.* (2000)
Bedrock channels	
• Formation of scabland	Starkel (1972)
• Channel bed and bank erosion	Kale (1990)
• Movement and deposition of coarse gravel	Kale *et al.* (1994)
in channel	Rajaguru *et al.* (1995)
• Building of gravel bars and boulder berms	Kale and Gadgil (1997)
in channel	Baker and Kale (1998)
• Erosion of gravel bars and boulder berms	Deodhar and Kale (1999)
• Deposition of sand and fines at tributary	Gupta *et al.* (1999)
mouths, bends and other protected areas as	Hire (2000)
slackwater deposits	

alaya (Dhar and Nandargi, 1998). Because of these meteorological situations and high relief, the Himalayan rivers frequently experience high flows and flash floods. During such events, considerable channel modification in the valleys takes place. Furthermore, in Himalaya, due to high rainfall intensity and unstable hillslopes mass movement processes are also important. Consequently, the stream channels display the combined effect of mass movements and floods.

Himalayan rivers flow thorough narrow and deep gorges and the channel gradients are at least an order of magnitude higher than the plains. Consequently, the hydraulic efficiency, the stream energy per unit area and the competence is much higher. Starkel (1972) described catastrophic changes in the Sikkim-Darjeeling Himalaya following the extraordinary rainfall-flood of October 1968. Besides rock and debris falls and slides, widespread incision and erosion was observed along major rivers, the Teesta and Great Rangit. Starkel (1972) has reported channel deepening up to 2–3 m in rock and an increase in channel width up to 80–200 m. The flood (unit discharges between 15 and 156 $m^3\ s^{-1}\ km^{-2}$; Gupta, 1988) was also associated with high competence and capacity. This was demonstrated by movement of large boulders up to 12 m in diameter. According to Starkel (1972) such channel-modifying events have a recurrence interval of 20–25 years in the Darjeeling Himalaya.

High-magnitude rainfall events in the Himalaya are also responsible for mass movement on slopes, which include landslides, rock/debris falls, avalanches and mudflows. Landslides are known to block rivers that cause massive floods when they fail. Numerous examples of breach floods have been recorded on many Himalayan rivers such as the Indus, Shyok, Lihut, Satluj, Bhagirathi, Alaknanda, Birehganga, Teesta and Subansiri (Hewitt, 1998; Wohl and Cenderelli, 1998; Shroder, 1998; Kale, 1998). During such events, even small streams demonstrate extreme competence and capacity. For example, rock mass measuring about a million cubic meters, dislodged during the August 1998 massive landslide, was washed away by a dam-failure flash flood on the Malpa Gad River in Kumaun Himalaya (Paul *et al.*, 2000). On some occasions, slope failures may be triggered by flood erosion. Sah and Mazari (1998) have reported undercutting of the unstable slopes by Beas River in spate in September 1995 that led to a major landslide in the Kulu Valley, Himachal Pradesh. The Alaknanada and Bhagirathi Rivers were similarly affected by landslide failure floods in 1970 and 1978 respectively.

4.1.2. *The Indus-Ganga-Brahmaputra Plains (Area II in Figure 1)*

Heavy rainfall over the IGB Plains is associated with cyclonic circulations and shift of the monsoon trough (Dhar and Nandargi, 1998). The eastern part of the Ganga and Brahmaputra Basins are subjected to large floods. On an average about 5 to 11 floods are recorded on different rivers of the IGB Plains every year. The duration of flood varies between 2 and 6 days.

As stated earlier, large floods on such rivers are associated with lower values of power per unit area, due to drastic width increase by overbank flows and/or channel widening. In addition, the width-depth ratio and power per unit area decreases during floods in some rivers, such as Brahmaputra, due to bed accretion (Richardson and Thorne, 2001). Low unit stream power values even for extreme flood events (Figure 4) suggest a general decrease in the ability to transport. This further implies that a large proportion of the sediments from the Himalaya Mountains is stored on the IGB plains. Goswami (1985) has estimated that in the Brahmaputra Basin about 70% of the suspended load is stored within the basin.

However, within the IGB plains, the morphologic properties and the response of rivers to floods vary depending on the type of river system. Sinha and Friend

(1994) have shown significant differences in the behavior of the rivers depending upon whether they are

(a) Mountain-fed.
(b) Plain-fed.
(c) Foothills-fed.

The mountain-fed streams originate in the Himalaya and have large basin areas in the mountains. The channels display braided patterns and the sediment load imposed from upstream is high. Kosi and Gandak are examples of such rivers. The mountain-fed rivers emerging onto the plains experience a rapid downstream decrease in unit stream power and shear stress, and hence are highly unstable (Kale, 1998). Further, due to large fluctuations in discharge and large quantities of bedload, the response to large floods is often dramatic. Such rivers build large fans and are known for their unusually high locational instability. The Kosi River, for example, has shifted by over a 100 km in the last 200 years or so (Gole and Chitale, 1966). In comparison, the plain-fed and foothills-fed rivers are highly sinuous and are not subjected to large variations in water and sediment discharge. Hence, the channels are relatively more stable (Sinha and Friend, 1994). This pattern of river behavior is, more or less, observed across the entire length of the IGB Plains.

The rivers of the Indus-Ganga-Brahmaputra Plains carry more than 2–3 orders of magnitude higher water and sediment discharge during floods than the non-monsoon low flows. The rivers have a wide floodplain on one or both the sides of the non-monsoon low flow channel. In the eastern parts of the Ganga Plains the floodwater inundate the floodplains at least once in two years (Sinha and Friend, 1994). The low flow channels frequently change their courses during monsoon floods, by avulsion or by bank cutting. It has been postulated that floods with a recurrence interval of 5 or more years are responsible for major changes in the courses of the rivers in the Ganga Plains (Asthana and Bhargava, 1981). Fluctuating stage and specific energy has been indicated as the primary cause of changes in the cross-sectional morphology of the Brahmaputra River (Richardson and Thorne, 2001).

Many workers have described the role of floods in avulsion and channel abandonment on the IGB plains and the alluvial fans. There are numerous reports of flood-induced changes in the channel sinuosity, length, size and shape in the Ganga-Brahmaputra Plains (Gole and Chitale, 1966; Goswami, 1985; Sarma and Basumallick, 1986; Singh, 1996; Sinha, 1996; Goswami *et al.*, 1999). In the eastern Ganga Plains (Bihar) Sinha (1996) noted that widespread overbank flooding and channel migration through avulsion and cut-off control the morphology

The stream channels on the IGB Plains are highly dynamic in nature and show frequent changes in the channel form during the monsoon season. Temporal changes in the channel position, bankline and length in the Ganga and Brahmaputra Rivers and some of their tributaries have been investigated by many workers. On the Brahmaputra River, Goswami (1985) reported an increase in the channel width by 3 km in 10 years (1971–1981). Studies by Basu *et al.* (1996) demonstrate that significant changes in the channel area of the Hugli River have occurred during 1973–1993. A study of bankline migration along the Burhi Dihing River conducted by Sarma and Basumallick (1984) revealed that the rate of bank erosion has varied between 2.4 and 3.5 km^2 (1919–1976). A similar study of the Subansiri River by Goswami *et al.* (1999) indicates that during 1920–1970 and 1970–1990 period about 107 and 57 km^2 area was respectively eroded. Godbole (1986) has estimated an increase in the channel length from 85 km in 1780 to 111 km in 1965–1966 of the Ganga River between Mokameh and Mansi. However, whether these changes reflect the cumulative effect of frequent, small-magnitude floods or the impact of a few large-magnitude floods remains uncertain and unclear. Some workers, such as Wells and Dorr (1987) and Sinha (1996) have attributed the changes in course of Kosi River and bank erosion in rivers of eastern Ganga Plains (Bihar) respectively to annual floods. Since large floods overtop the banks and increase the width-depth ratio, the shear stress and energy per unit area diminishes. Therefore, only those high flows that are confined within the banks and have relatively high bed shear stress appear to be geomorphologically more effective. Asthana and Bhargava (1981) are of the opinion that the floods with a magnitude of 0.5 to 0.8 of the maximum observed flood peaks (Q_{max}) are more effective and cause major changes in rivers courses in the Ganga Basin, and very high flood peaks are seldom responsible for major shifts in the river courses.

However, there are several instances that suggest that exceptional floods have caused significant changes in the river morphology and courses. There are reports of sudden changes in the river courses of the Beas River in 1943, the Yamuna River in 1924 and the Teesta River in the late 18th Century (Kale, 1998; and references therein). Sarma and Basumallick (1986) found evidence of bed erosion and scouring in the years with higher peaks and greater volumes of water discharge in Buri Dihing River, a tributary of the Brahmaputra River. Decrease in the channel length due to neck cut-off in Subansiri River, a tributary of Brahmaputra River in Assam, following the 1950 severe flood was inferred by Goswami *et al.* (1999).

These and a few other studies indicate that the river channel and floodplain morphologies in the IGB Plains are adjusted to the annual floods as well as occasional large floods.

4.1.3. *The Peninsular region (Area III in Figure 1)*

Floods on the Peninsular rivers are associated with active monsoon conditions over the Western Ghat and/or tropical cyclones originating over the Bay of Bengal and the Arabian Sea. Since the precipitation over the northern part of the Indian Peninsula is strongly concentrated in the summer monsoon season, the rivers in this part have a seasonal discharge related to the monsoon (Kale *et al.*, 1997). The rivers are generally incised in rock or alluvium and have stable channels. Changes in the bankline and shifts in channel position in response to floods are uncommon. It is only in the deltaic plains of Godavari, Krishna, Mahanadi and Kaveri Rivers, that sometimes changes in the channel form and position occur as a result of floods. The rivers usually display a channel-in-channel physiography, with high alluvial banks enclosing a small non-monsoon flow channel, patches of floodplain, and point bars (Gupta and Dutt, 1989; Gupta, 1995; Gupta *et al.*, 1999; Deodhar and Kale, 1999). Such channels experience overbank flows only on few occasions and, hence, all the fluvial activity is restricted within the high banks.

The role of large floods in such rivers has been assessed by evaluating the hydraulic characteristics of the flows by Kale *et al.* (1994), Gupta *et al.* (1999) and Deodhar and Kale (1999). These studies show that during large floods there is a significant decrease in the width-depth ratio and a corresponding increase in boundary shear stress and stream energy per unit area (Kale *et al.*, 1997). Since the incised channels are rarely filled up to the top, all large floods are associated with unit stream power values several orders of magnitude higher than the low non-monsoon flows. As a result, there are ample evidences of movement of coarse material on channel bed and changes in the channel bedform (Kale *et al.*, 1994; Gupta *et al.*, 1999; Deodhar and Kale, 1999). Multidate channel cross-section available for Tapi and some rivers of the Deccan Traps region show noteworthy changes in the channel bedforms, but marginal changes in the overall channel morphology (Deodhar and Kale, 1999; Hire, 2000).

Floods on the peninsular rivers are less frequent and generally the effects are less impressive. Consequently, there are very few studies on floods and their effects. Mujumdar *et al.* (1970) studied in detail the geomorphic impact of an extraordinary flood of September 1969, on the upper Godavari River. The peak discharge was close to 15,000 $m^3\ s^{-1}$ and the maximum velocities were in the range of 1.5–2.5 $m\ s^{-1}$. Unit stream power values were an order of magnitude higher than normal monsoon floods ($>110\ W\ m^{-2}$). Post-flood investigations indicate that the floodwater filled the entire channel, spread on both the banks, and deposited sandy-silt (20–50 cm) over the bank. This was unusual, because the channel is deeply incised and overbank flows are very rare. However, in spite of the high hydraulic radius, velocities and power per unit area, there were no notable changes in the channel size and morphology. Similarly, Rajaguru *et al.* (1995) did not observe any measurable change in the bankline and channel area after the August 1991 severe flood on the upper Narmada River near Jabalpur. Comparison of multidate cross-sectional data for the Tapi River also indicates that normal high flows are incapable of increasing the channel area or modifying the channel morphology (Hire, 2000). This fact implies that the channel resistance of the alluvial rivers in Peninsular India is usually not exceeded and the channel physiography is largely adjusted to extreme-magnitude floods that occur at large intervals of several decades or even longer.

The response of bedrock-influenced channels to large floods is different because of more resistant channel boundary conditions. This was demonstrated by a very large flood of July 1991, due to intense rainfall and a dam-failure on the upper Tapi River (Kale *et al.* 1994). Using a scourline carved by the catastrophic flood, the discharge was estimated to be about 9,000 to 10,000 $m^3\ s^{-1}$ (12 $m^3\ s^{-1}\ km^{-2}$).

The flood was associated with high unit stream power (ca. 1000 W m^{-2}) and was capable of erosion in rock and transportation of coarse sediments. Although about 2 m of soil was stripped from a large area, and large rock and concrete slabs were carried for a few tens of meters downstream, there were no significant changes in channel morphology or boulder bar morphology (Kale *et al.* 1994; Kale and Gadgil, 1997). Post-flood studies at 33 locations within the rocky gorge indicate that the flood effects include (Kale and Gadgil, 1997)

(a) Erosion of gravelly and sandy-silty terraces,
(b) Deposition of sand on terraces, bars and at tributary mouths,
(c) Bar reorganization, and
(d) Filling of deep pools by coarse sediments.

Bedrock erosion, scabland formation, and overbank gravel deposition was also observed at a few places.

The above examples indicate that the morphologic properties of the Peninsular rivers are adjusted to less frequent high-magnitude floods, because these are the only flows competent to alter the channel shape and size.

4.1.4. *The Thar Desert (Area IV in Figure 1)*

Floods in the Thar Desert of India are infrequent. However, sometimes heavy and widespread rains in association with monsoon depressions and low pressure systems generate rapid runoff and flash floods (Dhir *et al.*, 1982). Such a situation occurred in July 1979, when a massive flood occurred on the Luni River. Studies by Central Arid Zone Research Institute (CAZRI), Jodhpur, indicate that the peak discharge in the lower reaches ranged between 10,000 and 14,000 m^3 s^{-1} (Dhir *et al.*, 1982). As a result, the channel overflowed and the water spread up to a distance of 1–3 km on either side of the Luni River. The flood was responsible for spectacular changes in channel dimensions of the river in some reaches. Post-flood studies reveal an increase in the channel width up to half-a-km in some reaches (Dhir *et al.*, 1982; Sharma *et al.*, 1982). Sharma and Vangani (1992) found high suspended concentrations, ranging from 0.9 g/l in the foothills to 40.2 g/l at the basin outflow.

Figure 5 shows the geomorphic effects at different discharges. It can be seen from the figure that for low discharges, as expected, there was no erosion or sedimentation. However, it is interesting to note that the highest discharge (ca. 14,000 m^3 s^{-1}) close to the outflow was ineffective in terms of bank erosion as well as deposition (Figure 5). Maximum increase in the channel width as well as deposition was observed at sites that recorded discharges in the range of 5–7,000 m^3 s^{-1} (Sharma *et al.*, 1982). The reason for these differences in the geomorphic response is unclear. It is most likely that the associated flood stress and power was low at the outlet due to high width- depth ratio. This indicates that the absolute magnitude of the flow is not as important as the flood power, which is largely determined by the channel geometry (Kochel, 1988) and perimeter lithology (Baker and Kale, 1998).

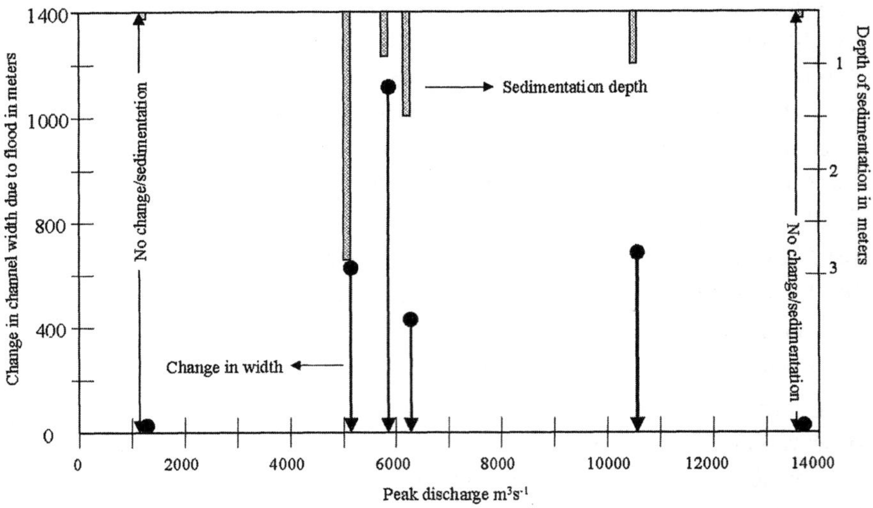

Figure 5. Diagram showing the increase in the channel width and the depth of sedimentation against peak discharge observed during the 1979 catastrophic flood on the Luni River. Basic data after Sharma *et al.* (1982).

Historical records suggest that this flood was the largest at least in the past 120 years (Dhir *et al.*, 1982). This fact implies that the high-magnitude floods play an extremely important role in shaping the channels in the semi-arid and arid regions and extreme floods create persistent flood related features.

5. Conclusion

Because the precipitation over a large part of India is strongly concentrated in the summer monsoon season during June and September/October, and the tropical storm season from May to October, almost all the geomorphic work is accomplished during the summer monsoon season. The river channels, consequently, are adjusted to the striking contrast between the discharge patterns of the monsoon and non-monsoon season, and periodic large floods.

Although all monsoon floods, big or small, have some geomorphic impact on the channel and floodplain morphology, the geomorphic effects of large floods are most impressive only in certain areas. These include (Kale, 1998) – (a) high-relief areas of the Himalaya, where channel gradients are at least an order of magnitude higher and bedload is dominated by coarse alluvium, and (b) alluvial channels in the arid and semi-arid areas, which are characterized by flash floods and easily erodible banks.

In the Indus-Ganga-Brahmaputra Plains the annual floods appear to be more important than occasional large flows in terms of avulsion, channel abandonment, and bank erosion (Asthana and Bhargava, 1981; Wells and Dorr, 1987; Sinha, 1996). In comparison, in the Indian Peninsula the river channels are more stable and respond only to extreme-magnitude floods that are capable of exceeding the channel resistance (even in alluvial channels).

The summary of several case studies presented above clearly indicates that it is not the absolute magnitude of discharge that is important in terms of geomorphic effects, rather it is the flood power that is largely determined by the channel geometry and perimeter lithology (Kale, 1998; Baker and Kale, 1998). Monsoon floods likely to result in significant geomorphic changes are those that produce high shear stress or stream power to entrain coarse sediments and to exceed the resistance provided by bed and banks.

Acknowledgements

This study was carried out in connection with several research projects sponsored by the Indian Department of Science and Technology. The author is grateful to Dilip K. Barua for constructive review of the manuscript.

References

Abbas, N. and Subramaninan, V.: 1984, Erosional and sediment transport in the Ganges River Basin (India), *J. Hydrol.* **69**, 173–182.

Asthana, B. N. and Bhargava, A. N.: 1981, Dominant discharge for alluvial rivers, *Irr. Power Jour.* **38**, 65–68.

Baker, V. R.: 1977, Stream channel response to floods with examples from central Texas, *Geol. Soc. Am. Bull.* **88**, 1057–1070.

Baker, V. R. and Costa, J. E.: 1987, Flood power, In: L. Mayer and D. Nash (eds), *Catastrophic Flooding*, Allen and Unwin, London, pp. 1–21.

Baker, V. R. and Kale, V. S.: 1998, The role of extreme floods in shaping bedrock channels, In: K. J. Tinkler and E. Wolh (eds), *Rivers over Rock: Fluvial Processes in Bedrock Channels*, American Geophysical Union Monograph, Vol. 107, pp. 153–165.

Basu, S. R., Sen, A., and Gosh, A.: 1996, A study on the morphological and hydrological changes of the Hugli River during 1973-1993', *Nat. Geog.* **31**, 57–68.

Bikshamaiah, G. and Subramanian, V.: 1980, Chemical and sediment mass transfer in the Godavari River Basin in India, *J. Hydrol.* **46**, 331–342.

Chakrapani, G. J. and Subramanian, V.: 1990, Factors controlling sediment discharge in the Mahanadi river basin, India, *J. Hydrol.* **117**, 169–185.

Coleman, J. M.: 1969, Brahmaputra River. Channel processes and sedimentation, *Sed. Geol.* **8**, 129–239.

Deodhar, L. A. and Kale, V. S.: 1999, Downstream adjustments in allochthonous rivers: Western Deccan Trap upland region, India, In A. J. Miller and A. Gupta (eds), *Varieties of Fluvial Form*, John Wiley and Sons, New York, pp. 295–315.

Dhar, O. N. and Nandargi, S. S.: 1993, The zones of severe rainstorm activity over India, *Int. J. Climatol.* **13**, 301–311.

Dhar, O. N. and Nandargi, S. S.: 1998, Floods in Indian rivers and their meteorological aspects, In: V. S. Kale (ed), *Flood Studies in India*, Geological Society of India, Bangalore, India, Memoir, Vol. 41, pp. 1–25.

Dhir, R. P., Kolarkar, A. S., Sharma, K. D., Vangani, N. S., Saxena, S. K, Sen, A. K., Ramakrishna, Y. S., Murthy, K. N. K., Singh, N., and Tak, B. L.: 1982, July 1979 Flash flood in the Luni. Jodhpur: Central Arid Zone Research Institute. Technical Bulletin No. 6.

Garde, R. J. and Kothyari, U. C.: 1990, Flood estimation in Indian catchments, *J. Hydrol.* **113**, 135–146.

Godbole, M. L.: 1986, Training of Ganga river from Mokamesh to Mansi, In: *Proceedings of the Seminar on Morphology of Ganga River*, Ganga Flood Control Commission, New Delhi, pp. 73–90.

Gole, C. V. and Chitale, S. V.: 1966, Inland delta-building activity of the Kosi River, *J. Hydra. Div. Am. Soc. Civ. Eng.* HY2, **92**, 111–126.

Goswami, D. C.: 1985, Brahmaputra River, Assam, India: Physiography, basin denudation, and channel aggradation, *Water Resourc. Res.* **21**, 959–978.

Goswami, D. C.: 1988, Magnitude and frequency of fluvial processes in the Brahmaputra Basin, Assam: Some observations, In: S. Singh and R. C. Tiwari (eds), *Geomorphology and Environment*, The Allahabad Geographical Society, Allahabad, India, pp. 203–211.

Goswami, D. C.: 1998, Fluvial regime and flood hydrology of the Brahmaputra River, Assam, In: V. S. Kale (ed), *Flood Studies in India*, Geological Society of India, Bangalore, India, Memoir, Vol. 41, pp. 53–76.

Goswami, U., Sarma, J. N., and Patgiri, A. D.: 1999, River channel changes of the Subansiri in Assam, India, *Geomorphology* **30**, 227–244.

Gupta, A.: 1988, Large floods as geomorphic events in the humid tropics, In: V. R. Baker, R. C. Kochel, and P. C. Patton (eds), *Flood Geomorphology*, Wiley, New York, pp. 151–177.

Gupta, A.: 1995, Magnitude, frequency, and special factors affecting channel form and processes in the seasonal tropics, In: J. E. Costa, A. J. Miller, K. W. Potter, and P. R. Wilcock (eds), *Natural and Anthropogenic Influences in Fluvial Geomorphology*, American Geophysical Union, Geophysical Monograph, Vol. 89, pp. 125–136.

Gupta, A.: 1998, Geomorphological effects of floods on Indian rivers, In: V. S. Kale (ed.), *Flood Studies in India*, Geological Society of India, Bangalore, India, Memoir, Vol. 41, 143–153.

Gupta, A. and Dutt, A.: 1989, The Auranga: Description of a tropical monsoon river, *Z. Geomorph.* NF **33**, 73–92.

Gupta, A., Kale, V. S., and Rajaguru, S. N.: 1999, The Narmada River, India, through space and time, In: A. J. Miller and A. Gupta (eds), *Varieties of Fluvial Form*, New York: John Wiley and Sons, pp. 113–143.

Hewitt, K.: 1998, Catastrophic landslides and their effects on the upper Indus streams, Karakoram Himalaya, northern Pakistan, *Geomorphology* **26**, 47–80.

Hire, P. S.: 2000, Geomorphic and hydrologic studies of floods in the Tapi Basin, Unpublished Ph.D. Thesis submitted to University of Pune, Pune, India.

Kale, V. S.: 1990, Morphological and hydrological characteristics of some allochthonous river channels, western Deccan Trap upland region, India, *Geomorphology* **3**, 31–43.

Kale, V. S.: 1998, Monsoon floods in India: A hydro-geomorphic perspective, In: V. S. Kale (ed), *Flood Studies in India*, Geological Society of India, Bangalore, India, Memoir, Vol. 41, pp. 229–256.

Kale, V. S.: 1999, Long-period fluctuations in monsoon floods in the Deccan Peninsular, India, *J. Geol. Soc. Ind.* **53**, 5–15.

Kale, V. S. and Gadgil, A.: 1997, Evaluation of the flood hydrology in the Upper Tapi Basin, Department of Science and Technology, Report.

Kale, V. S. and Gupta, A.: 2001, Introduction to Geomorphology, Orient Longman, Calcutta, 106 pp.

Kale, V. S., Ely, L. L., Enzel, Y. and Baker, V. R.: 1994, Geomorphic and hydrologic aspects of monsoon floods on the Narmada and Tapi Rivers in central India, *Geomorphology* **10**, 157–168.

Kale, V. S., Hire, P., and Baker, V. R.: 1997, Flood hydrology and geomorphology of monsoon-dominated rivers: The Indian Peninsula. *Water Inter.* **22**, 259–265.

Kochel, R. C.: 1988, Geomorphic impact of large floods: Review and new perspectives on magnitude and frequency, In: V. R. Baker, R. C. Kochel, and P. C. Patton (eds), *Flood Geomorphology*, John Wiley and Sons, New York, pp. 169–187.

Komar, P. D.: 1988, Sediment transport by floods, In: V. R. Baker, R. C., and P. C. Patton (eds), *Flood Geomorphology*, John Wiley and Sons New York. pp. 97–112.

Mujumdar, G. G., Rajaguru, S. N., and Papu, R. S.: 1970, The recent Godavari Flood (September 1969) and its relevance to prehistoric archeology, *Bull. Deccan College Res. Inst.* **24**, 1–17.

Paul, S. K., Bartarya, S. K., Rautela, P., and Mahajan, A. K. : 2000, Catastrophic mass movement of 1998 monsoons at Malpa in Kali Valley, Kumaun Himalaya, India, *Geomorphology* **35**, 169–180.

Rajaguru, S. N., Gupta, A., Kale, V. S., Mishra, S., Ganjoo, R. K., Ely, L. L., Enzel, Y., and Baker, V. R.: 1995, Channel form and processes of the flood-dominated Narmada River, India, *Earth Surf. Proc. Landform* **20**, 407–421.

Rakhecha, P. R. and Pisharoty, P. R.: 1996, Heavy rainfall during monsoon season: Point and spatial distribution, *Cur. Sci.* **71**, 179–186.

Ramaswamy, C.: 1987, Meteorological Aspects of Severe Floods in India 1923–1979, MMH No. 10, India Meteorological Department, New Delhi.

Ramesh, R. and Subramanian, V.: 1988, Temporal, spatial and size variation in the sediment transport in the Krishna River Basin, *India, J. Hydrol.* **98**, 53–65.

Richards, K. S.: 1999, The magnitude-frequency concept in fluvial geomorphology: A component of a degenerating research programme?, *Z. Geomorph.* SB **115**, 1–18.

Richardson, W. R. and Thorne, C. R.: 2001, Multiple thread flow and channel bifurcation in a braided river: Brahmaputra-Jamuna River, Bangladesh, *Geomorphology* **38**, 185–196.

Sah, M. P. and Mazari, R. K.: 1998, Anthropogenically accelerated mass movement, Kulu Valley, Himachal Pradesh, India, *Geomorphology* **26**, 123–138.

Sakthivadivel, R. and Raghupathy, A.: 1978, Frequency analysis of floods in some Indian Rivers, *Hydro. Rev.* **4**, 57–67.

Sarma J. N. and Basumallick, S. : 1984, Bankline migration of the Burhi Dihing River, Assam, *Ind. J. Earth Sci.* **11**, 199–206.

Sarma J. N. and Basumallick, S.: 1986, Channel form and process of the Burhi Dihing River, India, *Geograf. Ann.* **68A**, 373–381.

Sarma, J. N.: 1986, Sediment transport in the Burhi Dihing River, India, In R. F. Hadley (ed), Drainage Basin Sediment Delivery, IAHS Publication No. 159, pp. 199–215.

Sharma, K. D. and Vangani, N. S.: 1992, Characteristics of a flash flood in the Luni Basin, northwestern India, *Irr. Power Jour.* **49**, 31-41.

Sharma, K. D., Vangani, N. S., Chatterji, P. C., and Singh, G.: 1982, A severe flood in Luni Basin, western Rajasthan during July 1979 – A case study, *Mausum* **33**, 377–384.

Shroder, J. F.: 1998, Slope failure and denudation in the western Himalaya, *Geomorphology* **26**, 81–105.

Singh, I. B.: 1996, Geological Evolution of Ganga Plain – An overview, *J. Palaeonto. Soc. Ind.* **41**, 99–137.

Sinha, R.: 1996, Channel avulsion and floodplain structure in the Gandak-Kosi interfan, north Bihar plains, India, *Z. Geomorph.* **103**, 249–268.

Sinha, R. and Friend, P. F.: 1994, River systems and their sediment flux, Indo-Gangetic plains, Northern Bihar, India, *Sedimentology* **41**, 825–845.

Starkel, L.: 1972, The role of catastrophic rainfall in the shaping of the relief of the lower Himalaya (Darjeeling Hills), *Geog. Polon.* **21**, 103–160.

Starkel, L., Froehlich, W., and Soja, R.: 1998, Floods in Sikkim Himalaya – Their Causes, Curse and Effects, In: V. S. Kale (ed), *Flood Studies in India*, Geological Society of India, Bangalore, India, Memoir, Vol. 41, pp. 101–118.

Thatte, C. D., Rao, B. K., and More, D. K.: 1986, An approach to assess rate of bed load in large cobbly rivers. A case study of Narmada River, In: *Proceedings of the 53rd Research and Development Session*, Central Board of Irrigation and Power, New Delhi, India, pp. 237–251.

Wells, N. A. and Dorr, J. A.: 1987, Shifting of the Kosi River, northern India, *Geology* **15**, 204–207.

Williams, G. P.: 1983, Paleohydrological methods and some examples from Swedish fluvial environments. I. Cobble and boulder deposits, *Geograf. Ann.* **65A**, 227–243.

Wohl, E. E. and Cenderelli, D.: 1998, Flooding in the Himalaya Mountains, In: V. S. Kale (ed.), *Flood Studies in India*, Geological Society of India, Bangalore, India, Memoir, Vol. 41, pp. 77–99.

Natural Hazards **28:** 85–99, 2003.

The Disastrous Flood of 1998 and Long Term Mitigation Strategies for Dhaka City

I. M. FAISAL
Department of Environmental Studies, North South University, 12 Kemal Ataturk Avenue, Banani, Dhaka - 1213, Bangladesh
E-mail: imfaisal@yahoo.com

M. R. KABIR
Department of Water Resources Engineering, Bangladesh University of Engineering and Technology (BUET), Dhaka - 1000, Bangladesh
E-mail: mrkabir@wre.buet.edu.

A. NISHAT
IUCN-Bangladesh, House 3A, Road 15 (new), Dhanmondi R/A, Dhaka - 1209, Bangladesh
E-mail: iucn@citechco.net.

(Received: 28 June 2001; accepted: 4 March 2002)

Abstract. The disastrous flood of 1998 was a result of excessive rainfall all over the catchment areas of the major rivers of Bangladesh. Dhaka City, which is surrounded by rivers on all sides, was seriously affected despite the completion of Phase I of the Dhaka Integrated Flood Protection Project (DIFPP). Water entered into the protected part of the city through hydraulic leakage such as buried sewerage pipes, breached and incomplete floodwalls, ungated culverts and inoperative regulators. The drainage network and retention ponds of the city were found to be in poor conditions and capacities of the pumping stations were found inadequate. There was a serious lack of coordination between the agencies responsible for flood protection and drainage of the city. These issues must be addressed to achieve long-term flood mitigation. In addition, feedback from both the experts and general public indicated that completion of Phase II of DIFPP was essential to bring the eastern part of the city under flood protection. Other structural measures suggested in this paper include installing and maintaining adequate drainage and pumping capacity and timely operation of regulators. This study also suggests a set of non-structural measures for flood mitigation that include protecting the retention ponds, raising public awareness on maintaining the city drains, introducing land zoning and flood proofing in the eastern part of Dhaka, and stream lining institutional bottlenecks.

Key words: Dhaka, hydrology, flood management, structural measures, non-structural measures, land use plan, inter-agency coordination and multipurpose embankment

1. Introduction

Dhaka is the capital city of more than 120 million people of Bangladesh. Many of the important government offices, industrial headquarters, educational institutions, health facilities and private businesses of the country are located in Dhaka. Proper functioning of these institutions is vital for the growth and prosperity of the country. Dhaka City with a high population growth rate of 7% per year is expected to emerge as a mega city by 2010. Dhaka is surrounded by a network of rivers – the Turag on the west, the Buriganga on the south, the Balu on the east, and the Tongi Khal on the north (Figure 1). Water level in all these rivers rises in the monsoon (June–September) due to seasonal flooding and makes the city vulnerable to flood. In 1988, a major flood hit the country, including the capital city Dhaka, causing enormous damages to life, property and income. During this flood about 80% of the area under Dhaka City Corporation (DCC) went under water.

Consequently, a number of studies specifically addressing this issue have been carried out within the general framework of the Flood Action Plan (FAP) (FAP-8A and FAP-8B, 1991). These reports suggested a combination of structural and non-structural measures for flood protection and mitigation. As of 1998, Phase I of the recommendations including the Dhaka City Flood Protection Embankment along the Turag and the Buriganga rivers and improvement of the internal drainage system of the city, has been mostly completed. However, the eastern part of the city still remains unprotected. In 1998 a catastrophic flood hit Dhaka, which affected both the protected and the unprotected parts of the city due to the flood of 1998. During this flood, almost the entire eastern block (118 km^2) went under water and about 20% of the protected western part (136 km^2) was inundated due to hydraulic leakage through ungated drainage pipes and culverts at various locations. As a result, city life was severely affected and the government was under serious pressure to undertake long-term flood protection and mitigation measures to prevent such disasters in future. Box 1 summarizes damages to important infrastructure of the city due to the flood of 1998. Vulnerability of the city to floods is likely to increase in future due to the impacts of climate change initiated by the Greenhouse Effect. Experts fear that floods of magnitudes experienced in 1988 and 1998 may occur more frequently in future.

2. Objective and Methodology

The primary objective of this study is to develop a set of long-term strategies for and flood mitigation for Dhaka City. These strategies can be used as guidelines for flood management in other cities in Bangladesh and elsewhere.

The study has been carried out in a number of steps which include review of existing documents on past floods in Dhaka City, evaluation of current flood management practices, field visits, interview with key professionals, people's representatives and general public, and synthesis of the long-term flood mitigation strategies.

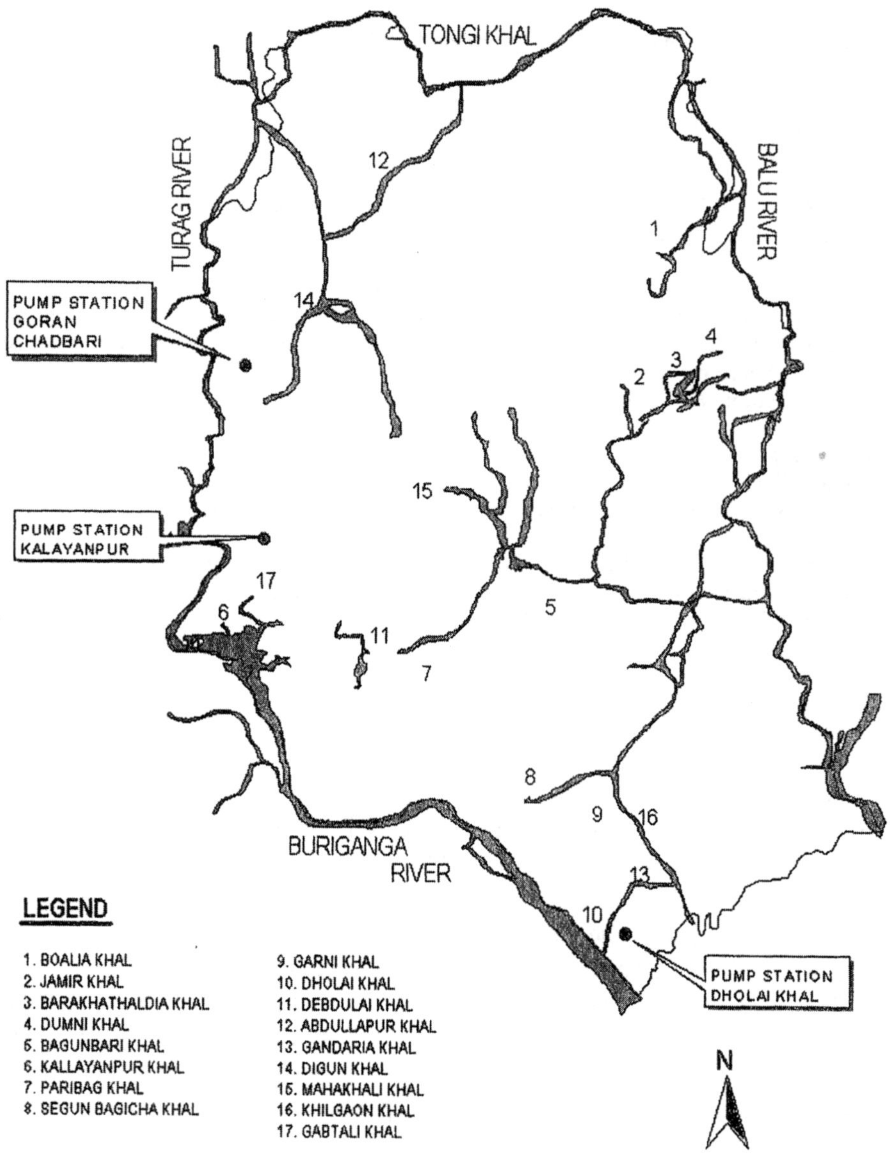

Figure 1. Rivers and canals of Dhaka City.

3. Floods in Dhaka

The historic cause of floods in Dhaka City is the rise in water levels of the rivers bordering the city. In addition, internal drainage congestion and in coordinated operation of flow regulation structures may make the flood situation worse. Figure 2 shows the pre-1988 inundation pattern of the city due to an average flood.

Box 1 Damages to important infrastructure of Dhaka City during the flood of 1998

***Housing*: According to a field survey carried by the Center for Urban Studies, Dhaka, approximately 262,000 houses of various types were damaged worth Taka 2.3 billion (Islam, 1998).**

***Roads*: About 1000 km of road surface was damaged within the DCC area (Siddique and Chowdhury, 1998).**

***Water supply*: According to the Dhaka Water and Sanitation Authority (DWASA), out of 235 deep tube wells, 12 went under water and had to be cleaned and restored after the flood. DWASA also had a plan to repair, clean and disinfect some 600 km of water supply mains. Additionally, 400 km of sewerage lines and 11 sewer lift pumps needed rehabilitation. Estimated cost of these activities was $10 million (Faisal, 1998).**

***Drainage*: A total of 223 km of box culvert, open drain and storm drain were reported to have silted up or damaged. About 510 km of tank sewer and sewerage line, 6 pump stations and 10 sewerage lift stations were affected (Siddique, 1998).**

***Power*: According to the Dhaka Electric Supply Authority (DESA), 500 transformers, 300 km of wire and 600 poles for 11 KV lines were damaged due to flood. Another 40 km of existing lines and 200 poles needed repair (Faisal, 1998).**

Box 1. Damages to important infrastructure of Dhaka City during the flood of 1998.

After the 1988 flood, the western and southern parts of the city have been brought under flood protection by constructing the embankment and floodwalls as per Phase I of Dhaka Integrated Flood Protection Project (DIFPP). However, these structures could not provide 100% flood protection in 1998 due to the presence of a few openings in the form of surface and buried drains and discontinuity in floodwalls. Figure 3 shows the areas that were inundated during the flood of 1998.

The eastern part of the city (to the east of Biswa Road) is not yet protected and all of it was inundated during the flood of 1998 as shown in Figure 3. This area largely comprises of low-lying floodplains that goes under water as the water level in the Balu River rises. However, growing population and land scarcity have forced many people to settle on this flood plain. Implementation of Phase II of DIFPP will provide flood protection to this part of the city.

4. Analysis of the Flood of 1998

4.1. PHYSICAL ASPECTS

In 1998, several floods passed through the country since early July. The flood situation started to worsen in the beginning of July due to heavy rainfall in the entire catchment area of the Ganges-Brahmaputra-Meghna system. Compared to the normal rainfall in the July–August period, the rainfall was 40% higher in July

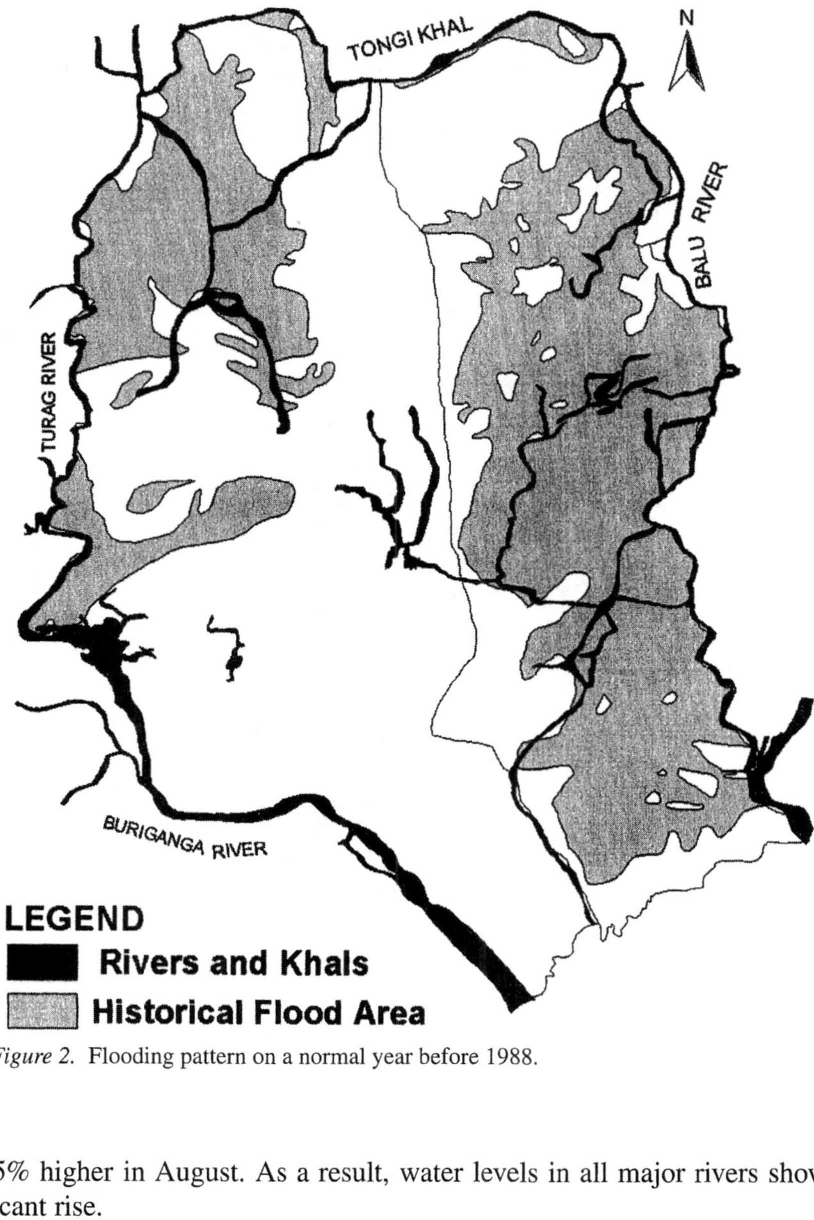

Figure 2. Flooding pattern on a normal year before 1988.

and 35% higher in August. As a result, water levels in all major rivers showed a significant rise.

The water levels in the rivers surrounding Dhaka approach the respective danger levels in the second half of July but cross the danger levels in mid-August. Peak flows occur typically in the last week of August and the first two weeks of September and flood water recedes in the last ten-days of September (Figures 4–7). The floods of 1987 and 1988 followed this pattern. However, during the flood of 1998, water levels crossed the danger levels almost a month earlier and stayed there until the last week of September. The main difference between the floods of 1988 and 1998 as seen in Figures 4–7, is that the former peaked very quickly in mid-August

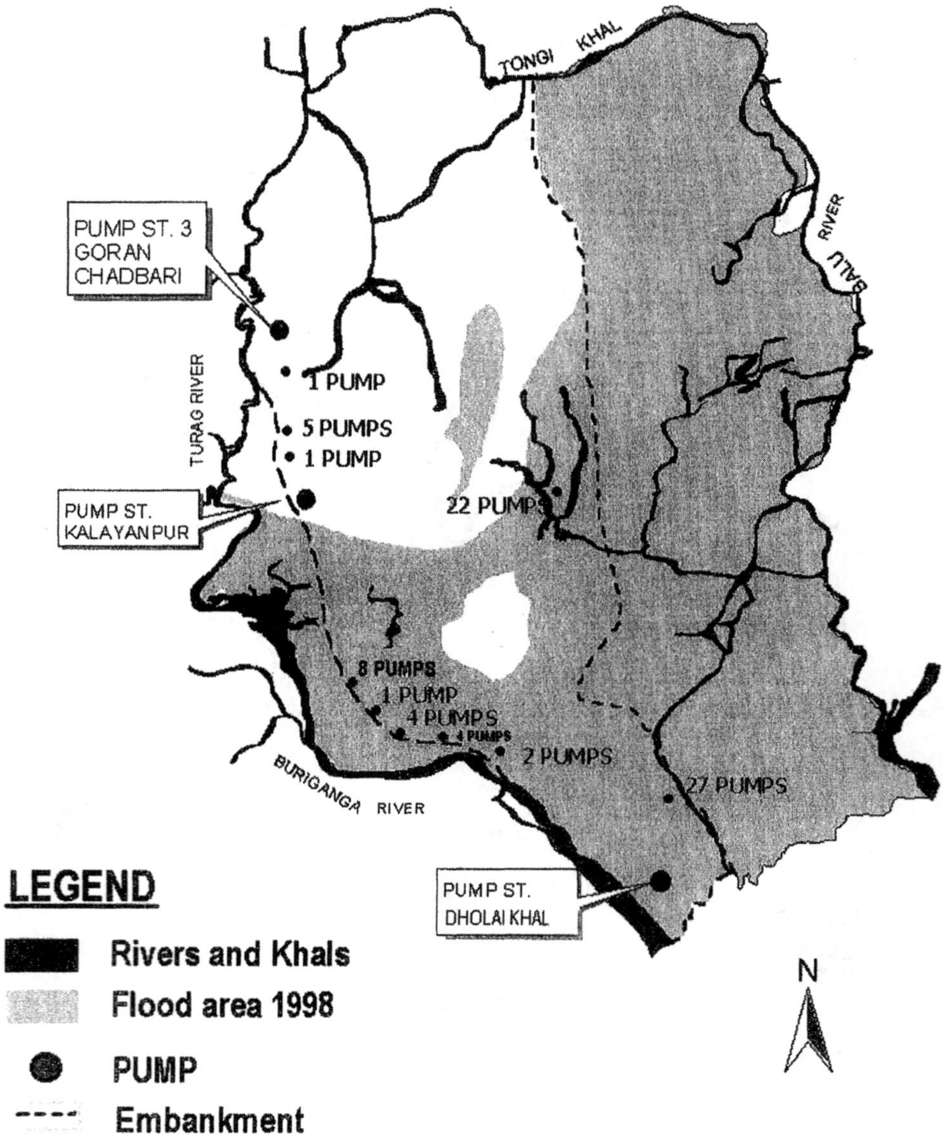

Figure 3. Inundation pattern in Dhaka City during the flood of 1998.

and had duration of about four weeks. The latter showed a smaller early peak, which prevailed until the arrival of a second peak in the second week of September. Finally the water levels receded in the last week of September. Thus the duration of 1998 flood was more than two months making it the longest flood in the history of Dhaka City. In terms of flood height, 1988 flood peaks were higher compared to the peaks of 1998.

Tongi Khal (Tongi)

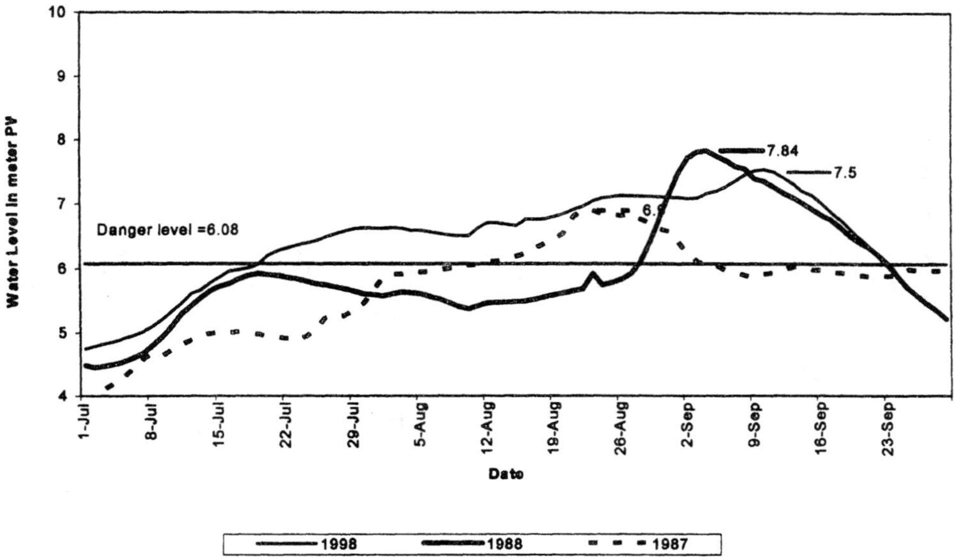

Figure 4.

Tongi Khal (Tongi)

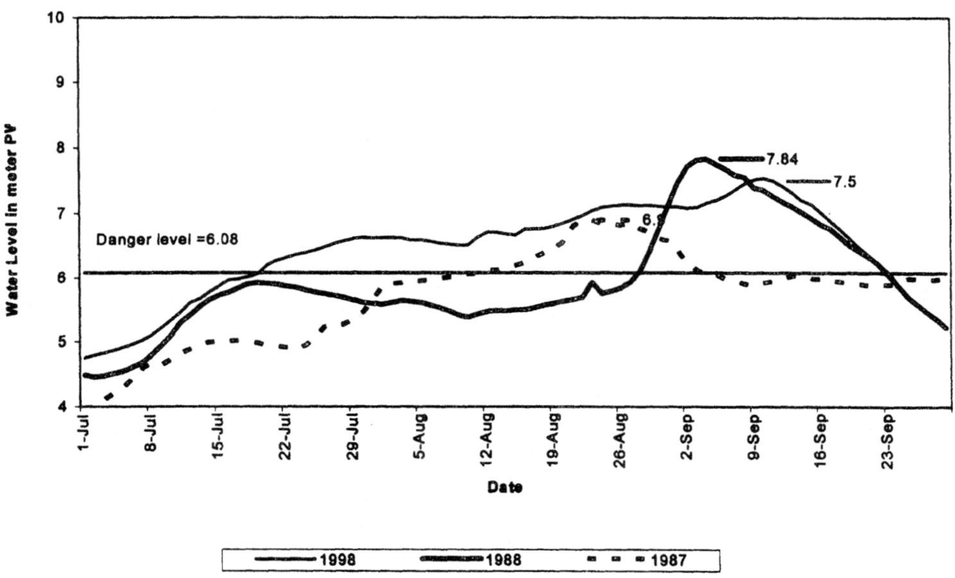

Figure 5.

Figures 4 and 5. Flood hydrographs for the Tongi Khal and the Turag River.

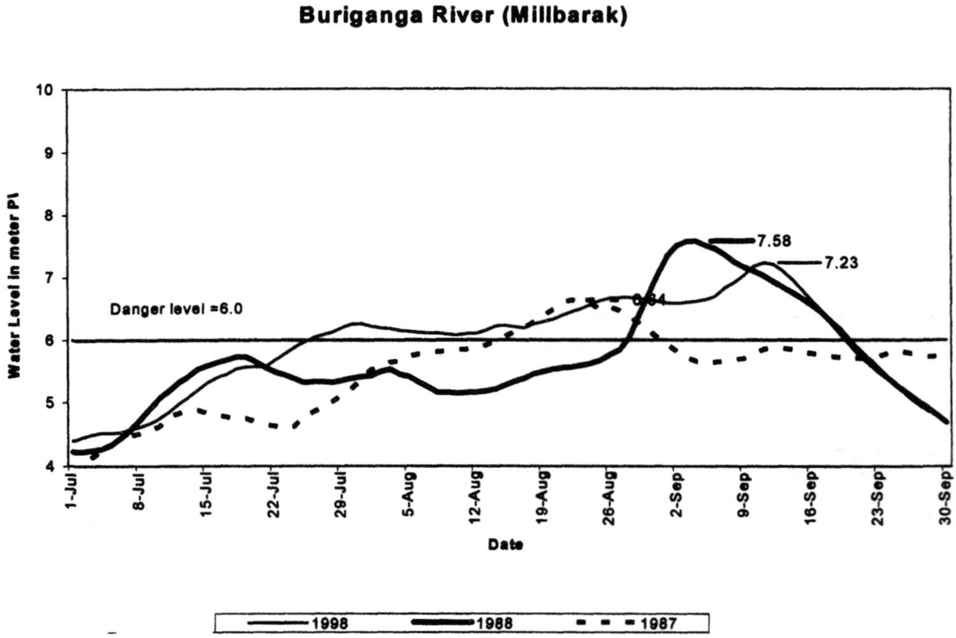

Figure 6.

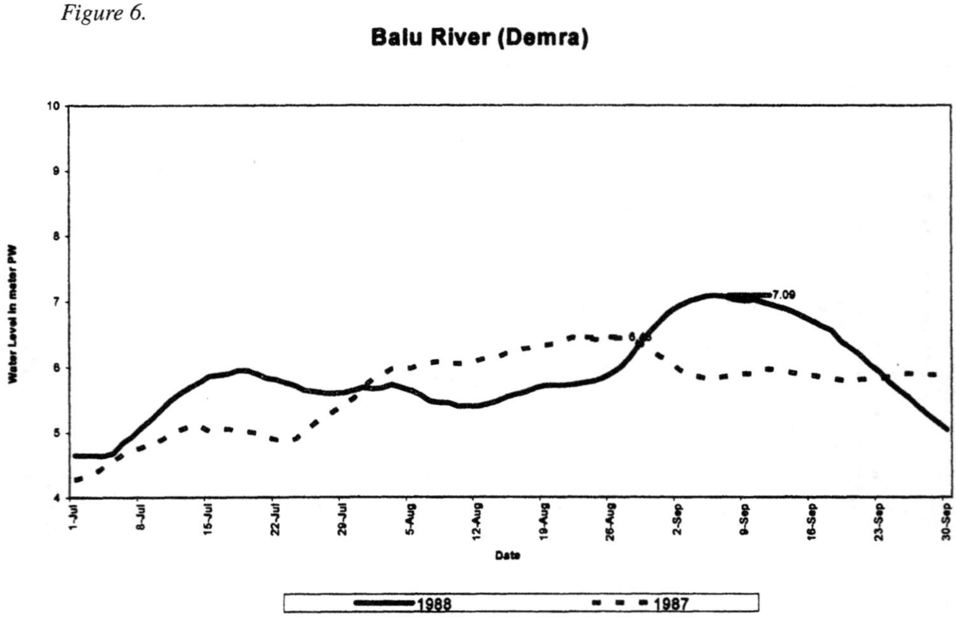

Figure 7.

Figures 4 and 5. Flood hydrographs for the Buriganga River and the Balu River.

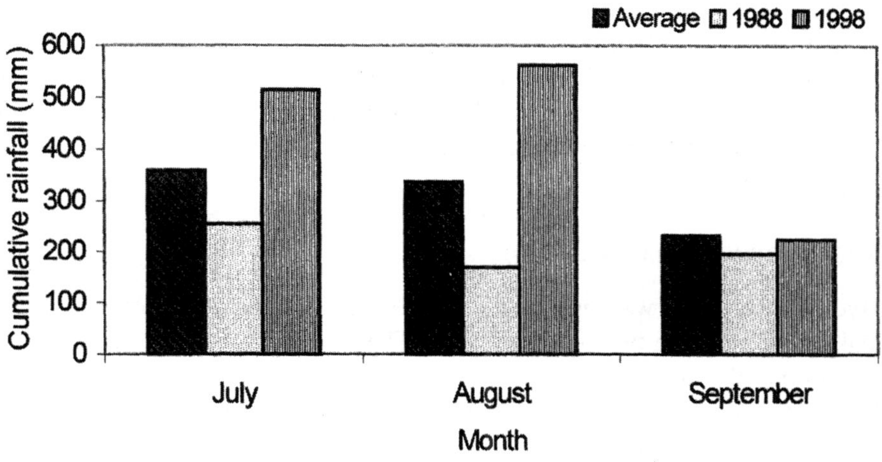

Figure 8. Cumulative monthly rainfall in the monsoon.

Another noticeable difference between the floods of 1988 and 1998 affecting Dhaka City is that the former was mainly caused by inflows from upstream areas, which caused a rise in the river water levels. Areas surrounding Dhaka City received less than average rainfall in 1988. In 1998, however, the rainfall was significantly higher than the average (Figure 8). This heavy seasonal rainfall and subsequent runoff might have contributed to the early crossing of the danger levels by all the rivers surrounding Dhaka City in 1998.

4.2. DRAINAGE CONGESTION AND INADEQUATE PUMPING

Heavy rainfall in the months of July and August of 1998 caused large parts of the city to become flooded. This drainage problem was caused by a number of factors. First, the existing drainage capacity was inadequate and in many places, the underground drains were partially silted up. Second, many of the retention ponds of the city have been encroached by developers leading to worsening of the problem. Currently the total area of retention within the urbanized part of DCC is about 900 acres. This area is less than 4% of the western part of Dhaka. According to the Rajdhani Unnayon Kotripokkhyo (RAJUK), about 12% of the urban area should be set aside as retention ponds (RAJUK, 1995). The third reason for internal flooding of the city in 1998 was that the water level in the Begunbari canal inside the city, which drains nearly 40% of the city, was lower than the water level of the Balu River since mid-July. Therefore, the sluice gate at Rampura, which connects the canal and the river, should have been closed and adequate pumping capacity should have been installed to prevent backflow from the Balu River and alleviate worsening drainage congestion. Unfortunately, due to the lack of inter-agency coordination, this could not be done in time.

Currently there are two pump stations operating in the western part of Dhaka. As per Phase I of DIFPP, one more station is under construction. Together these stations will have a total pumping capacity of about 70 m^3 s^{-1}. An additional 10 m^3 s^{-1} pumping capacity should be installed at Rampura regulator to prevent future drainage capacity problems as experienced in 1998.

4.3. HYDRAULIC LEAKAGE

Hydraulic leakage was one of the causes of flooding of the city in 1998. The leakage occurred in two different ways. The first one was due to incomplete construction of the floodwall along the Buriganga River in the southern part of the city. During implementation of Phase I of DIFPP, some existing structures could not be removed due to various reasons. One such prominent structure is the inland port of Dhaka at Sadarghat through which floodwater entered into the protected area. In a number of places, people removed parts of the floodwall for easy movement creating openings for floodwater. Moreover, floodwater back-flowed into the city through the sewerage pipes buried under the floodwall.

The second type of leakage occurred due to floodwater intrusion through open underground pipes and culverts in the northern part of the city. The Roads and Highways Department placed some of these pipes back in 1978 to ensure cross-drainage which were never monitored later on.

4.4. INSTITUTIONAL ASPECT

A number of organizations are responsible for flood protection and drainage of Dhaka City. The Bangladesh Water Development Board (BWDB) is in charge of flood mitigation for the entire country, including Dhaka. BWDB has the responsibility of constructing and maintaining all the embankments, floodwalls and regulators constructed under DIFPP. The Dhaka Water and Sanitation Authority (DWASA) is primarily responsible for drainage and operation of the pumping stations. A small part of the drainage network in the southern part of the city is maintained by DCC. In 1998, a lack of coordination among these organizations had contributed significantly towards prolonging the drainage congestion of the city. Specifically, the lack of coordination between DWASA and BWDB had contributed to a month-long delay in closing the regulator at Rampura Bridge during which floodwater entered into the city. Moreover, deployment of pumps was also delayed, as pumps had to be brought from another organization, the Bangladesh Agricultural Development Corporation (BADC). Many of these pumps were in the storage for nearly ten years (since the flood of 1988) and were inoperative. Pump operators complained that sufficient fuel was not allocated to operate these pumps on a continuous basis. Eventually, when the problems were taken care of by mid-August, a large part of the city was already inundated.

Table I. Summary of opinions of the key professionals

Question	BWDB	DWASA	DU	BUET
What are the main causes of the 1998 flood?	HR, PS, HT, TBF	HR, SM, Tsu	HR	HR, TBF
What do you think are the main causes of floods in Dhaka city?	IP, DC, HL	HL, GOF, IP	DC	DC, HL, GOF, IP
What major damages were caused by the 1998 flood to Dhaka city?	Rd, Em	Rd, H	H, Rd, I	Rd, H, I
How can Dhaka City be spared from floods in the future?	Phase I, II, TGO, AP, IDE	HL, TGO, AP, IDE, Phase II	Phase II	Phase I and II, IDE
What other things can be done to minimize flood damage in Dhaka city?	FW, FZ, IAC, FSh	PAA, IAC	FW, FSh	FZ, FSh, FW, IAC
How can the costs of flood protection/ control for Dhaka city be financed?	FA, SL, GOB	LDT, DSr	GOB	GOB, LDT

AP: adequate pumping; DC: drainage capacity; DSr: drainage surcharge; Em: embankment; FA: foreign aid; FSh: flood shelter; FW: forecasting and warning; FZ: flood zoning; GOB: government of Bangladesh; GOF: gate operation failure; H: housing; HL: hydraulic leakage; HR: heavy rainfall; HT: high tide; I: industry; IAC: inter-agency coordination; IDE: improved drainage efficiency; IP: inadequate pumping; LDT: land development tax; PAA: public awareness and activism; Phase I and II: phases of DIFPP; PS: Peak synchronization; Rd: road; SL: soft loan; SM: snow melt; TBF: transboundary flow; TGO: timely gate operation; Tsu: tsunami.

5. Feedback from Experts and the General Public

In order to incorporate views on long-term flood mitigation strategies from different sectors of the society, several key professionals, public representatives and a random sample of the general public were interviewed. The key professionals are with DWASA, BWDB, Bangladesh University of Engineering and Technology (BUET) and Dhaka University (DU). This group was involved in planning and implementation of various flood mitigation activities and their opinions are summarized in Table I.

Opinions expressed by the general public (one hundred randomly selected individuals divided in two groups – affected and non-affected) were similar. Most people identified heavy rain and trans-boundary flow as the primary cause of the flood of 1998 in Bangladesh. Exceedance of drainage capacity has been repetitively mentioned as the cause of flooding in Dhaka City. Road network has been mentioned as the most severely damaged sector, followed by water supply and housing. The people of Dhaka City unanimously believed that completion of DIFPP would spare the city from flooding in the long run. Flood proofing and improved flood warning systems have been mentioned as the most effective means of minimizing flood damage. As potential sources of funding for flood protection schemes, Government of Bangladesh (GOB) and foreign aid have been mentioned. A significant part of the respondents emphasized the need for local cost sharing. Interestingly more people from the non-affected group (47%) suggested cost sharing compared to the affected group (30%) as a means of funding. As a means of fund collection, special "flood tax" and lump sum fees (proportional to area) during land registration have also been suggested. Finally, individual preparedness has been mentioned as the best way to minimize personal loss due to flood. This reflects people's preference for self-reliance in combating the flood and the inadequacy of GOB and non-governmental organizations (NGO) measures in helping out the flood victims.

Ward commissioners of the flood-affected parts of the city were interviewed during the public opinion survey. This group was directly involved in flood fighting, relief and rehabilitation activities and was a better-informed group in terms of knowledge about flood damage mitigation. This group pointed out river siltation as the main cause of increased flood intensity. They emphasized the need to preserve the natural depressions and lakes within the city that serve as retention ponds for holding water from rain and internal drainage. They specifically mentioned that people living in slums had been badly affected by flood. This group suggested that more and better shelters could significantly reduce human sufferings and loss of life due to floods.

6. Major Findings from the Analysis of the 1998 Flood

Based on the analysis of the 1998 flood as presented above, a number of important findings are listed below in relation to flood mitigation in both the western and eastern parts of the city.

6.1. FINDINGS FOR THE WESTERN PART OF DHAKA

The western part of Dhaka City is largely protected by the embankment constructed after 1988 flood as per Phase I of DIFPP. Even then, about 20% of this part of the city was inundated in 1998 due to the following reasons:

i. Hydraulic leakage through un-gated drainage pipes and culverts;
ii. Delayed closure of the regulator gates at Rampura Bridge;
iii. Inadequate pumping from the retention storage;
iv. Leakage through incomplete segments of the floodwall along the Buriganga River;
v. Significantly reduced drainage capacity due to accumulation of solid waste and silt;
vi. Lack of coordination between organizations responsible for flood control and drainage;
vii. Partial completion of DIFPP.

It may be mentioned here that the embankment along the Turag River played a crucial role in protecting the western part of the city during 1998 flood even though its construction was not completed as per design. Specifically, the *berm* on the

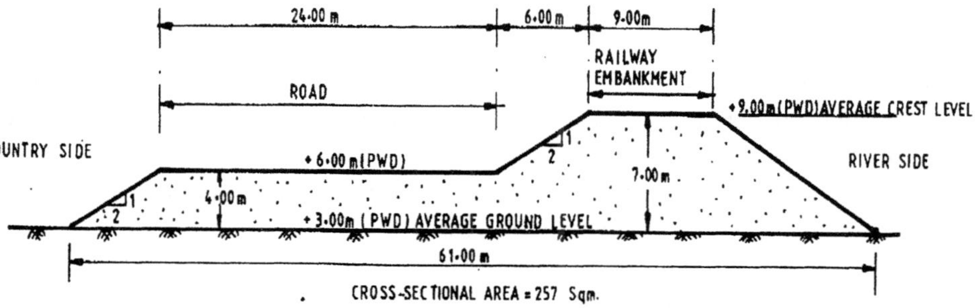

Figure 9. Cross-section of the multipurpose embankment along the Balu River.

countryside was not constructed and the crest width was found to be less than the design width in some parts of the embankment. As a result, there was a serious threat of seepage failure of the embankment, which was countered by holding water on the countryside with the help of sandbags.

6.2. FINDINGS FOR THE EASTERN PART OF DHAKA

The eastern part of Dhaka is flooded annually, as it is largely an unprotected low-lying flood plain. Anticipation of construction of the embankment along the Balu River had already triggered rapid urbanization in this part with little supervision from the city planning authority. Many structures are being built with low plinth level, which will be inundated in case of a major flood. The city planners must adopt and enforce proper land use plans in this area to prevent major economic loss and human sufferings in future. A very important aspect of the land use plan for Dhaka East would be the open areas to be developed as retention ponds to prevent massive water logging. Unless these areas are identified and preserved now, these will be taken up by urban developers and the city authority may have to buy these areas back at a much higher price compared to their present value.

7. Long-Term Strategies for Flood Mitigation

Some of the shortcomings of the current flood management practices have to be addressed immediately to protect Dhaka City from flooding. As short-term measures, all hydraulic leakage must be sealed off, badly silted-up drains must be cleaned and encroachment of retention ponds must be stopped. But to ensure long-term flood protection for Dhaka City, the most important step is to complete Phase II of DIFPP. After the 1998 flood, this plan has been expanded into a multi-objective project where the flood protection embankment will also house a four-lane road and a railway track (GOB, 1998). Figure 9 shows the proposed cross-section of the multipurpose embankment along the Balu River. The following structural and non-structural measures are recommended as long-term measures for flood mitigation.

7.1. STRUCTURAL MEASURES

i. Complete Phase II of DIFPP and ensure total flood protection for Dhaka City;
ii. Install and operate adequate number of pumps (supported by corresponding retention ponds) and maintain these in good condition;
iii. Maintain all existing and future flood control structures, reinforce weak points, provide bank protection where necessary;
iv. Maintain the internal drainage network in good condition; and,
v. Develop operating rules for all the regulators and establish close collaboration with the flood forecasting and warning agencies.

7.2. NON-STRUCTURAL MEASURES

i. Protect existing retention ponds in the western part of Dhaka and set aside sufficient land for retention ponds in the eastern part of Dhaka;
ii. Improve the flood forecasting and warning system, provide forecast in terms of potential area of inundation;
iii. Build public awareness on flood preparedness through NGOs and the media;
iv. Undertake awareness campaigns to prevent indiscriminate disposal of garbage in the surface and subsurface drains of the city;
v. Streamline institutional bottlenecks and assign all flood-related responsibilities to DWASA and BWDB as appropriate;
vi. Introduce rules and regulations for proper land use in the eastern part of Dhaka;
vii. Introduce mandatory flood proofing for new buildings in the eastern part of Dhaka;
viii. Provide flood warning and flood shelter;
ix. Ensure cost sharing by GOB and floodplain dwellers.

8. Conclusion

It is a national priority to protect Dhaka City from annual flooding, the impact of which is felt by the entire nation. A combination of structural and non-structural measures are suggested in this paper that will help achieve the goal of providing long-term flood mitigation for the city. Inter-agency co-ordination must be ensured for issuing timely flood warning and for operation and maintenance of the flood mitigation structures. For the eastern part of the city, proper land use plan must be introduced soon before the entire area is developed in an unplanned manner. People's feedback indicates that the rest of the country need not bear the entire cost of flood protection for Dhaka City. People living in the protected areas of the city are willing to share part of the cost through appropriately designed financial schemes.

References

Faisal, I. M.: 1998, *Damage Assessment of 1998 Flood*, Report submitted to the World Bank, Dhaka, Bangladesh.

FAP-8A: 1991, *JICA Master Plan Study for Greater Dhaka*, Japan International Cooperation Agency, Dhaka, Bangladesh.

FAP-8B: 1991, ADB *Dhaka Integrated Flood Protection Project (DIFPP)*, Louis Berger International, Inc., Dhaka Office.

GOB: 1998, *Dhaka Integrated Flood Control-cum-Eastern Bypass Road Multipurpose Project*, Working Paper (November 5, 1998) of the Project Steering Committee, Government of Bangladesh, Dhaka.

Islam, N.: 1998, *Flood '98 and the Future of Urban Settlements in Bangladesh*, National Seminar on Flood '98 and Management of Floods in Future (December 8, 1998), Jointly organized by National Committee for Relief and Rehabilitation and Dhaka University, Dhaka.

RAJUK: 1995, *Dhaka Metropolitan Development Plan (1995–2015)*, Vol. I & II, RAJUK, Dhaka.

Siddique, Q. I. and Chowdhury, M. H.: 1998, Flood-98: Losses and Damages, National Seminar on Flood '98 and Management of Floods in Future (December 8, 1998), Jointly organized by National Committee for Relief and Rehabilitation and Dhaka University, Dhaka.

Natural Hazards **28:** 101–129, 2003.

Floodplain Residents' Preferences for Water Level Management Options in Flood Control Projects in Bangladesh

HARUN RASID[1] and WOLFGANG HAIDER[2]
[1]*Department of Geography, Lakehead University, Thunder Bay, Ontario, Canada P7B 5E1;*
[2]*School of Resource and Environmental Management, Simon Fraser University, Burnaby, BC, Canada V5A 1S6*

(Received: 28 June 2001; accepted: 4 March 2002)

Abstract. A maximum difference conjoint (MDC) model was part of a questionnaire to assess floodplain residents' preferences for the outcomes of water level management inside an enclosed embankment system on the left bank of the Dhaleswari River, called the Compartmentalization Pilot Project (CPP)-Tangail. The outcomes were described as (a) varied flood depths in the rice field, (b) incidence of flooding on the courtyard and inside homes, (c) changes in floodplain and culture fisheries, and (d) varied conditions of drainage congestion. Following a standard survey, the respondents from inside and outside a completed section of the compartment (Cluster 1b) were shown profiles of these flood management outcomes and were asked to indicate for each profile the one most preferred and the one least preferred item. The results of the study indicated that the respondents had a clear preference for preventing flooding of their homes and courtyards and for an ideal water depth of 2 ft in the *aman* rice fields. At the same time, they also showed a strong concern about malfunctioning of sluices and to a lesser extent about the changes in the fish habitat. The successful application of the MDC as an approach to model tradeoffs among rural residents of Bangladesh shows that a relatively complex quantitative survey method, incorporating choice cards as pictograms, can be applied successfully even in a developing country.

Key words: compartmentalization project, regulated flood regimes, environmental impacts, floodplain residents' preferences, stated preference models

1. Introduction

Flood hazard research, which was pioneered by Gilbert White (1945 and 1974) and his followers (e.g., Kates and Burton, 1986; Palm, 1990; Smith, 1992; Burton *et al.*, 1993), has increased our understanding of floodplain residents' perceptions of flood problems as well as of their preferences for flood alleviation measures. However, in the past it had proven difficult to apply the results and methodologies of such studies explicitly in the decision-making process. The controversy over floodplain residents' preferences for proposed enclosed embankment systems in Bangladesh, called compartmentalization projects, provides a classic example of this difficulty. Successive surveys among the residents of major floodplains of Bangladesh by the

World Bank-sponsored Flood Action Plan (Dempster and Brammer, 1992) and by several independent scholars (such as Paul, 1995 and Rasid, 2000) indicated widespread support for regulated flooding by enclosed embankment systems. However, one of the shortcomings of this finding is that conventional survey-based research has been discredited because of the assumption that the researcher could manipulate the outcomes of a survey by predetermining the nature of response mode (Fischoff *et al.*, 1981, p. 19). A more fundamental shortcoming of hazard perception research is the discrepancy between the underlying theoretical assumptions of the *bounded rationality* and *economic satisficer* paradigm, and the methods used to research these cognitive processes. The bounded rationality and satisficing concepts view lay people as individuals with incomplete knowledge, striving to attain some satisfactory, though not necessarily maximal or optimal solutions (Slovic *et al.*, 1974. p. 189). The main problem is that most traditional research methods used to investigate people's perceptions and preferences fail to reveal more complex cognitive processes such as tradeoffs among the engineering, economic, social, cultural, and environmental dimensions of flood control projects. For example, potential environmental impacts of enclosed embankment systems in Bangladesh include loss of open-water fisheries, changes in aquatic ecology of the floodplain, and drainage congestion due to malfunctioning of sluices and culverts (Rasid and Mallik, 1995). The previous surveys, which indicated an overwhelming support for regulated flooding by enclosed embankments, did not evaluate any of these potential environmental tradeoffs of respondents' preferences for manipulated flood regimes. They evaluated the importance of or preference for its components singly, instead of incorporating the multivariate nature of the problem into the survey design explicitly.

The main objective of this paper is to compare floodplain residents' preferences for the outcomes of water level management inside and outside of a Compartmentalization Pilot Project (CPP-Tangail), by applying a multivariate stated preference approach, called maximum difference conjoint (MDC) analysis. A Rapid Rural Appraisal (RRA) inside and outside the CPP provided the basis for identifying the relevant environmental issues associated with water level regulation in this pilot project. Based on this appraisal, a set of MDC-based choice cards was developed and included in a standard questionnaire for eliciting floodplain residents' possible acceptance of management outcomes. By presenting complete profiles of outcomes, the survey considered the inevitable tradeoffs.

This paper is structured as follows. The results of a previous survey, conducted by the first author between 1991 and 1994 (Rasid, 2000), are summarized in the next section as a context for this 1996 experimental survey in the CPP. Following a brief description of the CPP-Tangail, the salient features of the RRA and the MDC model are presented. The first part of the section on results and interpretations presents floodplain residents' estimates of pre-project flood hydrology within their personal property and their assessments of positive and negative impacts of the CPP, including post-project changes in floodplain agriculture and fisheries. Thereafter, results of the multivariate logit regressions employed to analyze the MDC data are presented. We then conclude the paper with a discussion of implications of the findings to floodplain management, and with an evaluation of the merits of employing a method to investigate trade-off behavior explicitly.

2. Floodplain Residents' Preferences for Flood Alleviation Measures in Bangladesh: A Summary of 1991–1994 Survey

The flood control controversy in Bangladesh has evolved beyond some of the critics' simplistic assumptions of flood prevention by embankment. In the wake of the catastrophic floods of 1987 and 1988, the Government of Bangladesh adopted an elaborate scheme of flood alleviation measures under the umbrella of the World Bank-sponsored Flood Action Plan (FAP). The main structural component of the FAP is an innovative flood regulation scheme, which has been labeled as 'embankment Compartmentalization' (UNDP/GOB, 1989; World Bank, 1990; NEDECO, 1993). The objective of the scheme is not to prevent flooding, as it has been suggested erroneously by Custers (1992) and Haque and Zaman (1993), but to regulate flood levels inside the proposed flood-control compartments. This would allow the normal levels with which farmers are familiar, "excluding only the high, untimely or rapidly-rising floods, which are liable to damage crops and property" (Brammer, 1990). Despite this lofty goal, the FAP has faced mounting opposition from environmentalists (such as NGOs and academics), who are concerned about potential environmental impacts of the scheme (Boyce, 1990; Adnan, 1991; Adnan and Sufiyan, 1993; European Parliament, 1993; Rasid and Mallik, 1993, 1995; Hughes *et al.*, 1994).

Another major criticism of the FAP has been that it did not take into account the interests and views of the floodplain residents of Bangladesh (Adnan, 1991; Chowdhury, 1991; Rasid and Mallik, 1995). The 1991–1994 survey by the first author (Rasid, 2000) was designed mainly to study this specific issue, i.e., to obtain direct input of the floodplain residents regarding their choice of flood alleviation measures. More specifically, the study had two objectives. The first was to test the central hypothesis whether floodplain residents of Bangladesh preferred regulating flood levels in the floodplains (by enclosed embankment systems) as opposed to preventing flooding (i.e., by confining floodwater within embanked channels). The second objective was to assess compatibility of these preferences with floodplain residents' indigenous adjustments to floods as well as with the proposed compartmentalization scheme.

Figure 1 shows the approximate locations of the sixty-six sample villages, which were adjacent to the nearest fourteen towns (numbered in the map as 1–14). The numbers of respondents at each location are indicated in parenthesis. In all, nearly 1500 respondents were interviewed by a team of field assistants under the supervision of a field supervisor. Figure 2(a) presents the most relevant results of that study: a full 94% of respondents preferred regulated flood levels on the floodplain, instead of preventing flooding entirely by embankment (i.e., confining floodwater within embanked channels). The respondents were then asked to indicate their spontaneous preferences for flood depth in their crop fields, to assess compatibility of these preferences with the normal flood regimes. These responses have been classified into two ranges of flood depth in Figure 2(b). More than three-quarters (78%) preferred the range of 0.3 to 1.0 m (1 to 3.5 ft), which coincides with the ideal flood depths for transplant *aman* rice. The preference for deeper flood levels (1.1 to 2.5 m or 3.6 to 8 ft), on the other hand, seemed to coincide with the practice of inter-cropping deepwater broadcast *aman* and *aus* rice.

Further, to test for relationships between respondents' preferences for flood alleviation measures and their subjective evaluations of flood experience and flood characteristics, a number of multivariate logit /logistic regressions were performed. In these regressions, floodplain residents' preference for regulated flooding was used as the dependent variable and the respondents' perceptions of flood hydrology, flood losses, and indigenous adjustments to floods as independent variables (Table 1). Table 1(a) indicates that respondents who preferred regulated flooding experienced deeper flooding in their floating *aman* rice fields but lower depth of water in other *aman* rice fields. At the same time, they had recurrent normal annual floods in most of the crop fields. They were less likely to experience flooding on higher lands and were more likely to experience a shorter flood. Further, they were more likely to experience loss of jute during normal flood and loss of rice during the 1987 and 1988 floods. With regards to indigenous adjustments, respondents with greater preference for structural regulations of floods were more likely to have raised the barn for cattle, waded through flood water, used rafts for commuting, and cultivated transplant *aman* following initial flood damage to rice crops. They were less likely to have built platforms inside their homes or to have cultivated *aus* or jute on upper middle lands.

3. Compartmentalization Pilot Project (CPP-Tangail)

The shaded area in Figure 1 indicates the spatial extent of the FAP-proposed compartmentalization scheme. Had this scheme been implemented, eventually all major floodplains of Bangladesh would be converted into a series of contiguous flood-control compartments. However, the scheme is now on hold, pending the outcome of experimental studies in a number of pilot projects. The CPP-Tangail is a leading experimental project for monitoring its operational performance and post-constructional environmental impacts (Euroconsult *et al.*, 1995). The project is located on the left bank of the Dhaleswari, downstream of its confluence with the Brahmaputra (see Figure 1 for general location). The project is called CPP-Tangail because the flood control compartments surround the town of Tangail (Figure 3).

The construction of the CPP, which began in 1992–93, involved only limited amounts of upgrading and resectioning of a number of existing embankments and installing several drainage regulators (i.e., gated structures). The main embankment

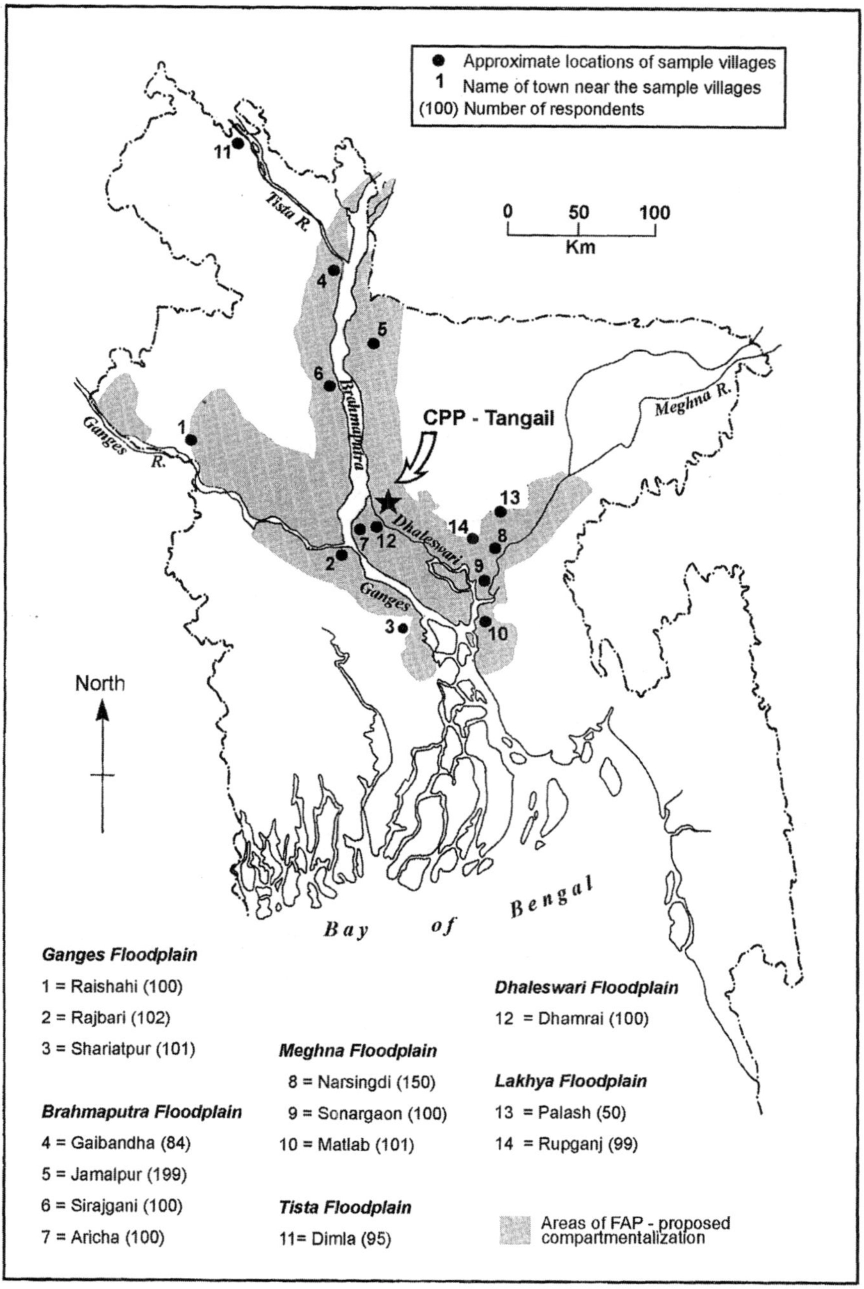

Figure 1. Map of Bangladesh showing approximate locations of sample villages for the 1991–1994 survey. Shaded areas indicate the extent of the FAP-proposed compartmentalization scheme.

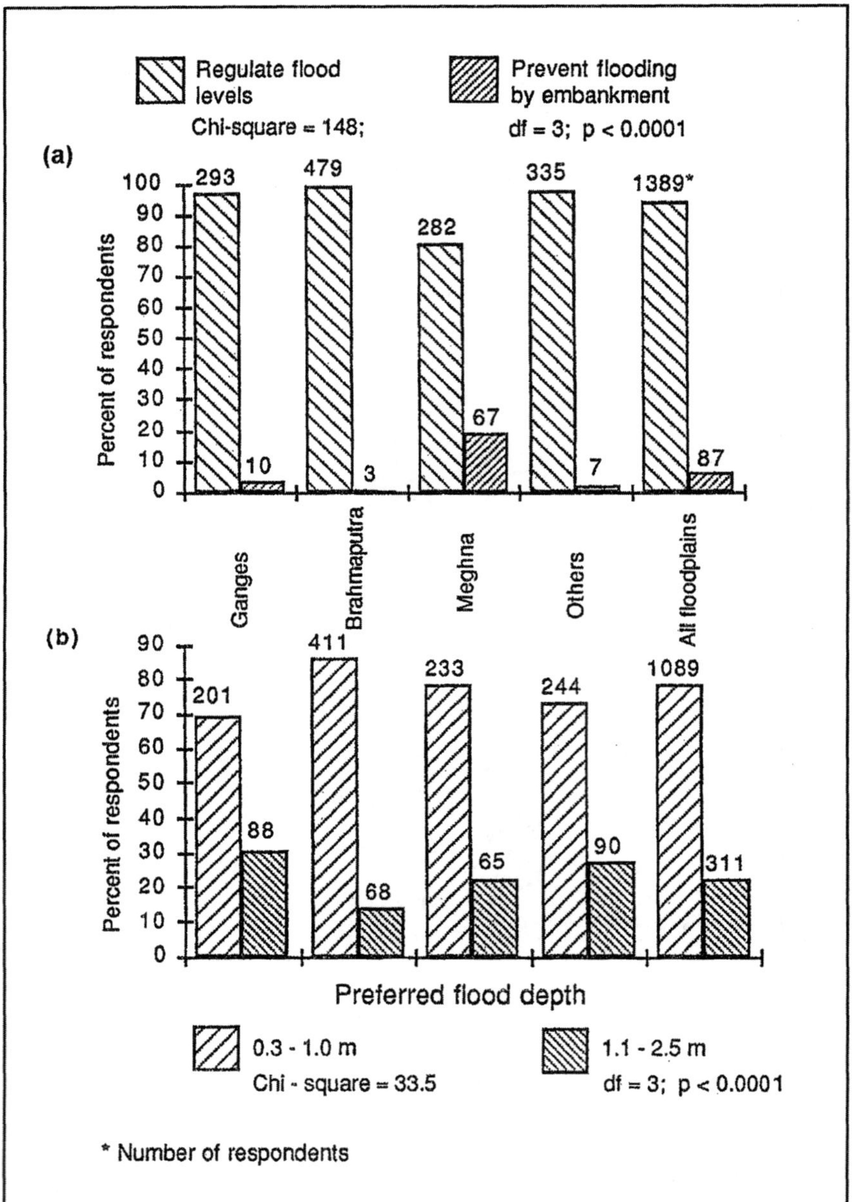

Figure 2. Summary of the 1991–1994 survey: (a) Floodplain residents' preferences for regulated flooding vs prevention of flooding by embankment, (b) ranges of preferred flood levels.

Table I. Logit/logistic regressions between floodplain residents' preferences for regulated flooding through compartmentalization and poldering in Bangladesh and selected explanatory variables

Independent variables	B	SE	p
(a) Perceived flood hydrology and flood losses			
Normal annual floods in crop fields (y/n)	1.7649	0.7162	0.0137
Depth of floodwater on high land (m)	−1.1247	0.4404	0.0107
Duration of normal floods (days)	−0.0136	0.0068	0.0434
Loss of jute during normal floods (%)	0.0145	0.0073	0.0469
Loss of rice in 1987–88 floods (%)	0.0527	0.0205	0.0100
Depth of floodwater in *aman* rice field (m)	−1.4451	0.5154	0.0050
Depth of floodwater in floating *aman* field (m)	2.3575	0.5226	0.0001
Constant = 2.33; −2 Log Likelihood = 146.78 (initial −2 Log Likelihood = 246.62) Model Chi-square = 99.84; df = 8; $p < 0.0001$			
(b) Indigenous adjustments to floods (y/n)			
Built platforms inside home	−1.8361	0.5204	0.0004
Raised barn floor for cattle	1.8727	0.3355	0.0001
Waded through flood water	1.6223	0.3718	0.0001
Used raft for commuting	3.6276	0.6314	0.0001
Cultivated *aus* on upper middle lands	−1.1632	0.3366	0.0005
Cultivated jute on upper middle lands	−1.6552	0.3775	0.0001
Cultivated transplant *aman* following flood damage of rice crops	3.2445	0.7674	0.0001
Constant = 2.11; −2 Log Likelihood = 279.75 (initial −2 Log Likelihood = 516.99) Model Chi-square = 237.25; df = 7; $p < 0.0001$			

forming the western, northern and eastern sides of the project was built in the mid-1970s under the United Nations Food for Work program. Similarly, a road embankment with open culverts and bridges forms the southern boundary of the project. The total area of the project (as shaded in Figure 3) is about 13,200 ha. It is further subdivided into a number of sub-compartments (bounded by lower embankments) and smaller field size water management units called *chawks* (bounded by village roads and settlement areas) (Euroconsult *et al.*, 1995, p. 8). At the time of this study (July–October 1996), only the northwestern section of the project, called Cluster 1b (with an area of 2447 ha), was completed (see Figure 3). Inside this cluster there are three sub-compartments and 39 *chawks*.

The main embankments forming the compartments of CPP-Tangail can provide limited protection against floods with return periods of up to 12 years (Euroconsult *et al.*, 1995, p. 11). The first test of Cluster 1b occurred during the 1995 summer

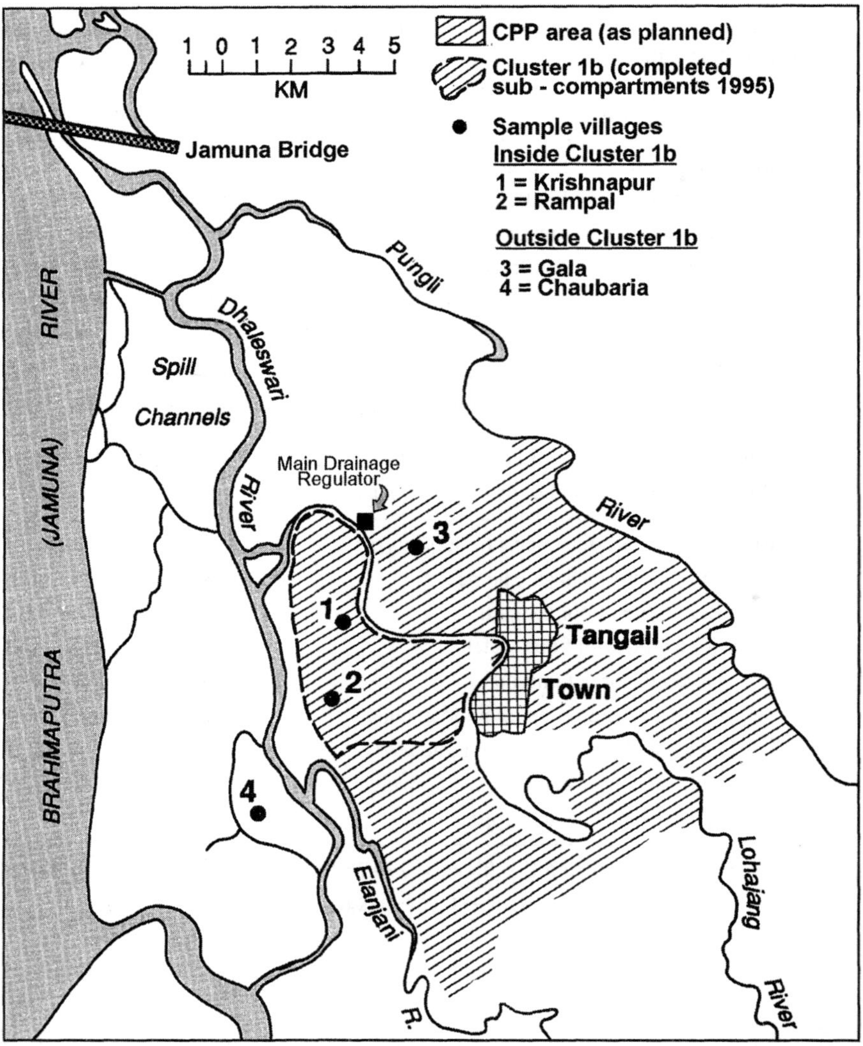

Figure 3. Location of the CPP-Tangail.

monsoon, when the peak water level in the Brahmaputra was only 40 cm short of the historic 1987 level. This flood amounted to only a partial performance testing of the cluster because not all planned engineering interventions were in place yet. In particular, the main drainage regulator along the northern embankment was not yet installed with gates. Despite these shortcomings, water levels inside the completed cluster were much lower than they would have been without the project. At least two negative impacts were also observed. First, a small area (up to 150 ha) immediately upstream of the main regulator experienced increased water levels, ranging from 10 cm to 50 cm, due to the throttling effect of the regulator (Euroconsult *et al.*, 1995, pp. 70–71). Second, there was also a significant decline in the open-water fish catches inside the cluster (Euroconsult *et al.*, 1995, p. 56).

4. Methodology

4.1. RAPID RURAL APPRAISAL (RRA)

Rapid Rural Appraisal and Participatory Rural Appraisal have been proposed as alternative methodologies for research in rural communities, in response to the critique that conventional questionnaires are extractive in nature and are incapable of representing the concerns of the affected population adequately (Whyte, 1986). These approaches enable local people to share, enhance, and analyze their knowledge of local conditions (Chambers, 1994). Ideally, participatory research calls for sustained collaboration between the researchers and local residents. Because of the logistic constraints, it was not feasible for the current authors to undertake a truly participatory appraisal. Instead, an RRA approach was adopted, in which a field supervisor and his field assistant made several reconnaissance trips to sample villages. The following RRA activities were performed.

Taped interviews. Unstructured interviews were conducted with the local residents in four sample villages for about five hours. Topics of conversation included (a) local flood regimes in the CPP area, (b) cropping calendar of the area, (c) relationships between flood hydrology (regimes, depths, etc) and sowing/harvesting regimes of different varieties of rice, (d) relationships between flood levels and rural infrastructures, including homesteads, housing structures, roads and embankments, etc. The respondents were also asked to describe their experiences of the 1987 and 1988 floods. They were further encouraged to comment on the positive and negative aspects of embankments. Did they want embankment? Why or why not? Did they experience any problem with the performance of the CPP?

A *transect walk* involved sketching the profile (relative height) of the land above the surrounding water levels in the floodplain along a path from one end of a village (*bari*) to another. The field supervisor recruited a number of volunteers to help him in this task. The sketched transects included relative elevations of rural roads, embankments, homesteads, houses on the courtyard, etc in relation to the water levels in adjacent low-lying areas (floodplains, rice field, etc).

Field sketches of rice fields were prepared, based on reconnaissance surveys in a selected number of shallow and deepwater plots. Varied depths of water in these plots were measured by using a pole from a country boat.

4.2. STATED PREFERENCE MODELS

Although the field level appraisals, such as an RRA, is an appropriate approach for obtaining insights into floodplain residents' concerns and opinions about flood alleviation measures and various policy alternatives, these methods are not capable of accommodating the multi-attribute nature of floodplain management problems formally. In the case of the CPP, the floodplain management decisions would require an evaluation of potential impacts of regulated flood regimes on various environmental attributes of the floodplain, such as the housing structure, rice cultivation, open-water and culture fisheries, and drainage conditions prior to the implementation of the project. Many of the impacts of water level regulation in the CPP, as presented in this paper, are only postulated. Whenever decision-making processes involve outcomes of hypothetical alternatives, stated preference models enable researchers to ask respondents for an evaluation of such non-existent alternatives.

One fundamental distinction in the stated preference research is between compositional and decompositional approaches, depending on whether only single components of a policy or full profiles of policy alternatives are evaluated by a respondent (Timmermanns, 1984). The latter is accomplished through decompositional approaches, in which alternatives are defined as combinations of a set of attributes, and each set is evaluated as a whole. Researchers then decompose these overall evaluations through statistical analysis and thus derive part-worth utility functions for each of the single components of these alternatives. If these full profile evaluations are carried out on each profile separately in the form of a rating or ranking task, the technique is usually referred to as conjoint analysis. In a major methodological advance, Louviere and Woodworth (1983) suggested combining two or several hypothetical profiles in choice sets, from which the respondents would then choose the most preferred alternatives. The resulting responses enable the researcher to base the analysis on the discrete choice model (McFadden 1974; Ben Akiva and Lerman, 1985); hence conjoint-based choice models are frequently referred to as discrete choice experiments. A variation to the choice experiment is the maximum difference conjoint (MDC) analysis, in which individual profiles are constructed in an identical manner to conjoint analysis (Louviere *et al.*, 1993). However, only one profile is presented to respondents at any one time, and instead of evaluating the profile as a whole respondents are asked to identify the "one most preferred" and the "one least preferred" feature (attribute level) in each profile. Cognitively this response task involves an evaluation (or judgement) of the highest utility difference between the two selected attributes out of a choice of set of attributes (i.e., the profile). In other words, a respondent's joint choice of the "most" and the "least" attractive feature of each profile can be viewed as a report of the two attribute levels that are furthest apart on an underlying utility scale for a given profile of attribute levels (Louviere *et al.*, 1993).

The analysis of the MDC is typically based on the discrete choice model (McFadden, 1974), which argues that choices can be modeled as a function of the attributes of the alternatives relevant to a given choice problem. It is assumed that the part-worth utilities associated with each of the attributes are integrated cognitively into overall utilities of the alternatives, and the alternative with the highest overall utility is selected. The processes of evaluation by respondents consist of two parts: a deterministic component that describes the formal utility, which the consumer derives from the alternative, and a stochastic component that describes the error. The error may account for omitted explanatory variables, individual variations in taste, and measurement error. This relationship is typically formulated as:

$$U_j = V_j + e_j, \tag{1}$$

where U_j is the utility of alternative j, V_j is the deterministic utility of alternative j, and e_j is the associated error term. If one assumes that, for the entire sample, the stochastic elements of the utilities follow a Gumbel distribution, the standard multinomial logit (MNL) model can be specified (McFadden 1974; Ben-Akiva and Lerman 1985):

$$\text{Prob}\{i \text{ chosen} = e^{Vi} / \sum e^{Vj}, \tag{2}$$

where the aggregate probability of choosing alternative i equals the exponent of all the measurable elements of alternative i over the sum of the exponent of all measurable elements of all j alternatives. This standard multinomial logit model supports the estimation of parameters that allow one to express the choice probability of a given alternative as a function of the attributes comprising that alternative and those attributes of all other alternatives in the choice set.

The discrete choice model can be used as the statistical method of analysis of an MDC, because formally the difference between the two attribute levels can be expressed in the following manner (Louviere *et al.*, 1993):

$$D_{ij} = \eth_{ij} + e_{ij}, \tag{3}$$

where D_{ij} is the true, but unobservable difference between attribute levels i and j; $\eth_{ij}$ represents the utility scale difference between levels i and j of the two different attributes; and e_{ij} is a random component associated with that difference. If the respondent reports the two levels that are furthest apart, one can estimate the probability that a particular pair ij is the largest or maximum in all possible comparisons of pairs in a set of features (C):

$$P_{ij}/C = P[(\eth_{ij} + e_{ij}) > (\eth_{kl} + e_{kl})], \quad Vkl = ij \in C. \tag{4}$$

The analogy of this formulation to the discrete choice model defined above should be obvious, only in this case the model describes the process of joint choice of two objects which are furthest apart on some underlying dimension of interest (i.e., utility). The empirical frequencies for each of the possible combination of pairs of the "most preferred/least preferred" features in each profile constitute the dependent variable. In this way one can derive an interval utility scale, where one attribute level sets the point of origin, and differences between that level and all other attribute levels on the underlying scale can be estimated from the choice

Table II. Experimental variables and their levels

Variable (attribute)	Attribute levels
Incidence of flooding on the homestead	(a) Neither house nor yard flooded
	(b) Yard flooded
	(c) House and yard flooded
Depth of floodwater in the *ama* rice field	(a) 2 ft
	(b) 5 feet
	(c) 8 feet
Changes in floodplain and culture fish	(a) Decrease in floodplain and culture fish
	(b) Decrease in floodplain fish but increase in culture fish
	(c) No change in floodplain or culture fish
Changes in drainage condition	(a) Stagnation of water due to malfunctioning of sluices
	(b) No problem with water drainage

data. For an application of the MDC method in resource management, see Haider and Hunt (1997).

4.3. APPLICATION OF THE MDC MODEL IN THE SURVEY

The survey had two components. The first part was a standard questionnaire, which was designed to collect three types of descriptive data. These included basic demographic and socioeconomic data on the respondents; their perceptions of pre-project hydrology and post-project environmental impacts of the CPP (including changes in flood hydrology, cropping practices, and floodplain fisheries); and a 5-point Likert scale for rating their perceptions of some of the postulated positive and negative impacts of the CPP.

The second part of the survey instrument contained the MDC experiment, consisting of a set of response profiles. The hypothetical profile on each card consisted of four variables (also called attributes) and each variable was specified on either two or three levels (Table II). These attributes were selected from the relevant literature and as a result of the RRA in the sample villages. Two of the four variables dealt with flood regimes, and two with environmental effects. All variables were treated as categorical in nature.

Following a randomized incomplete block design, nine scenarios (choice cards) were sufficient for estimating all of the main effects (Cochran and Cox, 1968). To facilitate explanation of the task, each of the choice cards was prepared as a pictogram and was laminated so that each could be presented separately to the respondents as a flash card (Figure 4). A tenth card containing a scenario that was

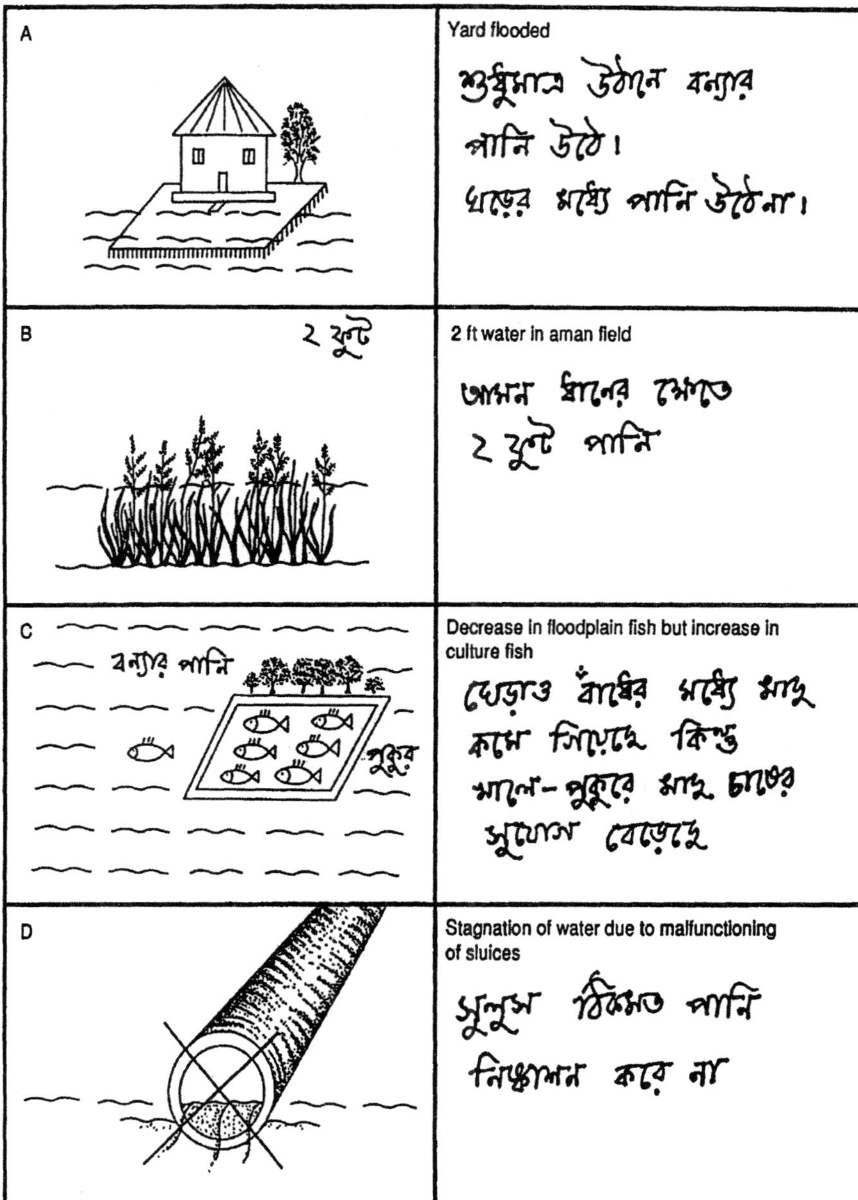

Figure 4. Sample choice card (pictogram) used in the MDC questionnaire survey.

not part of the design was presented as the first card to familiarize the respondent with the task. Typically each interview (including both the descriptive part and the MDC) lasted between 30 and 45 minutes.

4.4. THE SAMPLE

Of primary interest in the current study were the opinions and preferences of floodplain residents, who were presumably impacted by the CPP. Following the RRA, two pairs of villages were selected: one pair from inside and the other from outside Cluster 1b. In each pair, one of the villages was situated in a shallow or moderately deep flood zone; the other represented a deep flood zone. Although Gala (village 3) is shown in Figure 3 within the CPP planned area, at the time of the survey it was outside Cluster 1b (i.e., outside the completed section of the CPP). Chaubaria (village 4), situated on a low-lying alluvial island (*char*) between the Dhalweswari and the Brahmaputra, is truly outside the project.

Given comparable physical settings of the sample villages (i.e., shallow and deep flood zones) and similar socioeconomic characteristics of the population, a systematic sampling procedure was adopted to obtain representative samples from both inside and outside Cluster 1b. A typical village in the CPP area, as in most other areas of Bangladesh, consists of a number of homesteads (*bari*). The latter is an elevated tract of land – a natural levee or an artificially raised mound – on which a group of residents (normally kin) build their households (living quarters and other attached structures). Individual households were the basic sampling units. The target was to sample 50 households from each of the four villages to obtain a total sample of 200.

Starting approximately from the centre of a sample village, all of the households situated on adjacent homesteads were counted and numbered on a sketch, until the number reached 200 to 250. Then, using a systematic sampling procedure, every fourth or fifth household (depending on the size of the community) was selected into the sampling frame, until a sample of 50 was achieved for each village. The male household heads were selected as respondents. The females were excluded because of the restrictions of *purdah*, which prohibited contact with strangers (Rasid and Mallik, 1993). A field supervisor conducted a door-to-door interview with the help of a field assistant. If no one was at home, or if a potential respondent was unable or unwilling to have an interview, the next household was selected as the replacement sample.

5. Results and Interpretations

5.1. DEMOGRAPHIC AND SOCIOECONOMIC CHARACTERISTICS OF THE RESPONDENTS

The majority of the respondents were between 41 and 60 years of age, with a mean age of 44.7 years (standard deviation = 13). Most of them were long-term residents at their inherited property, with a mean tenure of 43.2 years (SD = 13.1). The majority of the respondents were subsistence farmers, with relatively small plots of lands (median size = 100 decimals). Almost all of the respondents sold surplus rice in the market and at least 50% earned between Taka 5,000 and 15,000 per year from the sale of rice (US $1 = approximately Taka 50). Besides rice, nearly three-quarters sold jute as a cash crop. About one-third supplemented their income by working as day labourer. There were no significant differences among the four villages in most of these variables, with the exception of the size of land holdings. Respondents from inside Cluster 1b (i.e., Krishnapur and Rampal) had significantly larger land holdings than those from outside.

5.2. RESPONDENTS' ESTIMATES OF PRE-PROJECT FLOOD HYDROLOGY

Using White's (1974) flood hazard perception paradigm, one of the central assumptions of this study was that the floodplain residents' experience and perceptions of past flood hazards would influence their preferences for water level regulation inside the CPP. Therefore, respondents were asked to provide estimates of flood hydrology (such as depth and frequency of flooding) in different components of their personal property. Even though such data might be considered as tentative, they provided useful estimates of floodplain residents' adjustments to normal flood regimes. Elevations of the homesteads above the surrounding floodplain and of the plinth levels of individual homes above the courtyard constituted the most basic adjustments to floods (Rasid and Paul, 1987; Rasid, 2000).

An overwhelming 72% of the sampled homes were of medium quality, with earth foundation/floor and corrugated tin roof. About 18% were of poor quality construction, with earth foundation /floor and thatched roof. The remaining 10% were well-to-do homes with cement floor and corrugated tin roof. Only a limited few of the dwellings were brick buildings. There were significant differences in the quality of housing between one of the sample villages inside Cluster 1b and another from outside: all dwellings in Krishnapur (inside) were of medium and good quality, compared to only 68% in Gala (outside).

Interval scale data on elevation of the homestead and depth of flooding were obtained from spontaneous responses to open-ended questions. Mean elevation of the courtyard (i.e., homestead) above the floodplain ranged between 1.34 m and 1.55 m. As the ANOVA tests in Table III indicate, mean elevations of homesteads in Krishnapur and Rampal (inside villages) were significantly higher than those in Gala and Chaubaria (outside villages). However, there were no significant spatial variations in elevations of the plinth levels above the courtyard, with mean elevations ranging between 0.45 m and 0.51 m.

Since the compartmentalization project is intended to regulate high magnitude flood levels, respondents were asked to report the depth of flooding on their courtyards as well as inside their homes during the catastrophic floods of 1987 and 1988. There were significant variations in these estimated depths from one sample village to another, but pairwise comparisons in ANOVA and Kruskal-Wallis H tests indicated that these differences did not display any consistent inside/outside patterns. Thus, the mean depth of flooding on the courtyard was 0.8 m in Krishnapur (inside) and 0.73 m in Chaubaria (outside), compared to 0.6 m in Rampal (inside) and 0.64

Table III. Respondents' estimates of pre-project flood hydrology: inside vs outside Cluster 1b

Items assessed	Inside Cluster 1b		Outside Cluster 1b		
Village	1	2	3	4	
				Pairwise difference	
Elevation of courtyard above the floodplain (m)				N = 200*	$p < 0.0001$
Mean	1.55	1.49	1.34	1.37	1 and 2
Standard deviation (SD)	0.27	0.27	0.24	0.21	from 3 and 4
Elevation of the plinth level above courtyard (m)				N = 199*	$p < 0.433$
Mean	0.48	0.48	0.45	0.51	
SD	0.12	0.15	0.15	0.12	
Depth of flooding on the courtyard during 1987–88 floods (m)				N = 193*	$p < 0.0001$
Mean	0.85	0.6	0.64	0.73	1 and 4
SD				from 2	
					1 from 3
Depth of flooding inside homes during 1987-88 floods (m)				N = 130**	$p < 0.006$
Mean	0.48	0.24	0.33	0.36	1 and 4
Median	0.3	0.3	0.3	0.3	from 2
Depth of flooding in the *aman* rice field during 1987–88 floods (m)				N = 199*	$p < 0.0001$
Mean	2.95	2.46	2.92	3.2	1, 3 and 4
Std dev	0.54	0.36	0.6	0.42	from 2
					3 from 4

*ANOVA tests (Tukey's Honestly Significant Difference pairwise t-tests).
**Kruskal-Wallis H tests (Bonferronni adjusted U-tests for pairwise comparisons).

m in Gala (outside) (Table III). The mean depth of flooding inside homes, ranging between 0.24 m and 0.48 m, displayed a similar inconsistent pattern of inside /outside variations. Although damage to personal property from such inundation was obviously high, no damage data were sought because of the methodological problems of estimating such damage.

Besides damage to property, the record-high magnitudes of the 1987 and 1988 floods caused massive losses to the *aman* variety of rice, which is the most important monsoon crop in Bangladesh (Brammer, 1990; Johnson, 1982). Because of such massive crop losses from two successive high-magnitude floods, protection of monsoon crops by preventing high-magnitude floods became the main priority of the FAP-proposed compartmentalization scheme (Rasid and Mallik, 1995, p. 8). For that reason, respondents were asked to estimate the depth of flooding specifically in their *aman* rice field. One of the assumptions of such a specific question was that if the depth of flooding in respondents' *aman* rice field were "too high" during the 1987 and 1988 floods, these data would then provide some references for more desirable flood depths inside the CPP. As data in Table III indicate, the mean depth of flooding in the *aman* rice field was estimated between 2.46 m and

3.2 m. Indeed, these depths might be considered as too high since they surpassed the critical depth of 2.4 m (8 feet), which has been estimated as the tolerable limit for even the deepwater varieties of *aman* (Paul, 1984; Rasid and Paul, 1987; Rasid and Mallik, 1995). There were significant variations in depth of flooding in the *aman* rice field from one village to another, but again there were no consistent pairwise inside/outside differences.

Besides the depth of flooding, estimates were also sought for the frequency of flooding on the homestead, from a set of close-ended responses. Thus, the respondents were asked to estimate the frequency of flooding on the courtyard of their homesteads and inside their homes from three predetermined ranges of return periods, i.e., every year, once in 2–5 years, and once in >5 years. These ranges were suggested to the respondents to control their responses within more familiar hydrological terms for return periods and to avoid misinterpretation of the time frames by different respondents (Rasid, 2000, p. 53). The data on perceived frequency of flooding (not reported in any table here) indicated that most of the homesteads were well adjusted to recurrent annual floods. None of the respondents reported flooding every year inside their homes and only four respondents from Krishnapur (village 1) reported inundation of their courtyard every year. An overwhelming 94% reported flooding inside their homes only during high magnitude flood events, such as the 1987 and 1988 floods. Similarly, 88% reported inundation of their courtyards once in >5 years. There were no significant differences among the sample villages in these variables.

5.3. RESPONDENTS' ASSESSMENTS OF POST-PROJECT CHANGES IN FLOODPLAIN AGRICULTURE AND FISHERY

Table IV summarises responses about post-project changes in floodplain agriculture and fishery. In addition to respondents' perceptions of post-CPP changes in these variables, these responses were most likely also influenced by the outcomes of deliberate testing and monitoring of changes by the consultants during implementation of the CPP project. Traditionally Tangail is an area of jute and deepwater *aman* cultivation (Johnson, 1982; Euroconsult *et al.*, 1995). The CPP testing programs included a shift from some of the deepwater *aman* to local transplant *aman* and a major expansion of the area under high yielding varieties of rice (Euroconsult *et al.*, 1995, p. 49). These changes required regulated shallower water levels in the rice field, which were achieved by the CPP testing program with considerable success. An increase in the HYV rice area resulted in a corresponding decrease in *aus* and jute cultivation.

The respondents were asked to indicate their perceptions of some of these changes by using a 3-point Likert scale. Most of the responses confirmed the expected changes almost in a predictable manner. Thus, there was a significant increase in *aman* cultivation (including the transplant variety) inside Cluster 1b, whereas outside the cluster, in Gala and Chaubaria, the increases were much more

Table IV. Respondents' assessments of post-project changes in floodplain agriculture and fishery: inside vs outside pairwise comparisons (data in percent of respondents)

Items assessed	Inside Cluster 1b		Outside Cluster 1b		
Village	1	2	3	4	
Pairwise difference					
Pairwise difference Rice cultivation: *aman* variety					
Decrease	0	8	16	28	N = 200
Same	44	16	72	70	$p < 0.0001$
Increase	56	76	12	2	
Rice cultivation: *aus* variety					
Decrease	74	66	34	36	N = 200
Same	22	30	66	60	$p < 0.0001$
Increase	4	4	0	4	
Rice cultivation: HYV *IRRI* variety					
Decrease	0	0	0	0	N = 200
Same	46	66	64	62	$p < 0.0001$
Increase	54	34	36	38	
Jute cultivation					
Decrease	54	60	30	26	N = 200
Same	38	36	66	58	$p < 0.0001$
Increase	8	4	4	12	
Quantity of floodplain fish					
Decrease	90	100	42	48	N =200
Same	2	0	56	50	$p < 0.0001$
Increase	8	0	2	2	
Quantity of culture fish					
Decrease	0	0	10	0	N = 200
Same	8	2	18	8	$p < 0.0001$
Increase	92	98	72	92	

modest. More striking were the significant decreases in *aus* cultivation inside the cluster: three-quarters of the respondents from Krishnapur and two-thirds from Rampal reported such decreases. Outside the cluster, also about one-third of the respondents from both Gala and Chaubaria reported decreases in *aus* cultivation, but these changes were more likely related to the general expansion of the HYV IRRI in Tangail area at the expense of other crops. For the same reason, at least one-third from Gala and Chaubaria and 34–54% from Krishnapur and Rampal reported increases in IRRI cultivation. The expansion of HYV rice has also occurred at the expense of jute. More than one half of the respondents from inside Cluster 1b reported decreases in jute cultivation, compared to 26–30% outside.

Fish is the most important source of animal protein in Bangladesh. The floodplain residents of the Tangail area depend heavily on the open water fishery of the floodplain during the monsoon season. Nearly two-thirds of the respondents have been fishing in the floodplain on a regular basis. Throughout the remaining periods of the year, especially during the dry season, standing water in marshes (*beels*) and other depressed parts of the floodplain provide the main fishing grounds. At least 70% reported fishing in such depressions. About one in every two has also been fishing occasionally in nearby rivers and ponds.

The impact of regulated water levels on different aspects of floodplain fishery constituted the second most important objective of the CPP testing and monitoring programs. The results of most of these tests were inconclusive by the end of the 1995 monsoon season, mainly because of incomplete engineering interventions in Cluster 1b. However, limited data on catch rates in one of the *beels* inside Cluster 1b indicated a significant decrease in catch with reductions in water depth in this *beel* (Euroconsult *et al.*, 1995, p. 55). In an attempt to compensate for the loss of capture fisheries due to the intervention of the CPP, an aquaculture extension program had been initiated throughout the entire CPP area. About 80% of the ponds of the area had been stocked with culture fish, prior to the present survey. Respondents' estimates of changes in floodplain and culture fish, thus, reflected the negative impacts of intervention by the CPP as well as the positive effects of aquaculture extension program. Reported decreases in open-water fisheries on the floodplain suggest a major loss of the fishery inside Cluster 1b: 90–100% of the respondents reported these decreases. Although these reported decreases were significantly higher inside the cluster, the intervention apparently has also affected the capture fisheries of the surrounding floodplain, as 42-48% of the respondents from Gala and Chaubaria also reported decreases in floodplain fish. In contrast, culture fisheries have increased everywhere: 92–98% reported such increases inside the cluster, compared to 72–92% outside.

5.4. RESPONDENTS' ASSESSMENTS OF POSITIVE AND NEGATIVE IMPACTS OF THE CPP

To gain further insights into respondents' perceptions of potential environmental impacts of the CPP, they were asked to rate four statements describing positive and negative impacts of the CPP project on a five-point Likert scale (Table V). Respondents from both of the communities inside Cluster 1b and those from Gala (outside) rated the effectiveness of the project for flood alleviation relatively high (mean scores = 3.5–3.7) As the ANOVA pairwise t-tests indicate, the respondents of Chaubaria rated this effect significantly lower (mean score = 2.8). The different assessment in Chaubaria can be explained easily, as this community is situated on a low-lying alluvial island (*char*) of the Dhaleswari River outside the CPP and was not exposed to the regulated flood regimes of the CPP. However, the more favourable ranking of this item by the respondents from Gala (mean score = 3.5) is somewhat surprising, in view of their experience with deeper flooding in certain sections of this village, due to the backwater effect of the main drainage regulator of the CPP. Although this assessment seemed to support the consultants' assertion that only a small area of Gala (up to 150 ha) had been affected by this backwater effect, a very high mean score of 4.0 for worsening of flood outside the project contradicted this assumption. Despite experiencing deeper flooding outside the project, it was entirely plausible that the residents of this village, being adjacent to the project, were exposed to the efficacy of engineering intervention and were, thus, convinced that the enclosed embankment system of the CPP was an effective measure for regulating flood levels. Transcripts of taped interviews confirmed this assumption. Despite persuasive attempts by the field supervisor and his field assistant to point out some of the adverse impacts of embankments, many of the participants in the taped conversations insisted on the need for embankment in their village. In general, the mean scores for worsening of flood outside the project displayed a more consistent pairwise inside/outside difference. Predictably the respondents from outside Cluster 1b had significantly higher scores for this adverse impact of the project (mean scores = 3.1–4.0) than those from inside (mean scores = 2.1–2.2). The persistent emphasis on embankment apparently reflects the floodplain residents' perception that this measure was the only effective method of protection. Therefore we considered it important to assess these opinions by focusing on an evaluation of outcomes and on various tradeoffs that were inevitable between alternatives.

Reduction of floodplain fishing due to the engineering intervention by the CPP has been assessed in an inconsistent manner. There was no apparent reason for higher rating of this item by the respondents from Chaubaria, unless it could be assumed that the project had cutoff the migration routes of open-water floodplain species. Although monitoring of catch rates inside Cluster 1b had indicated a significant post-project decline in open-water fishing (Euroconsult *et al.*, 1995, p. 56), respondents' assessments of this problem seemed to be somewhat restrained (with mean scores of 2.5–3.5). Finally, residents of the area seemed to have relatively low concerns about drainage congestion in enclosed embankment systems. Mean scores of 1.6–1.9 for this item meant that they were either not at all concerned or somewhat concerned about this issue.

5.5. RESULTS OF THE MDC EXPERIMENTS

One of the central assumptions of this study was that the floodplain residents of the CPP area would be willing to accept some of the adverse environmental impacts of enclosed embankment systems as tradeoffs for the benefits of regulated flood regimes. An alternative assumption was that if the adverse impacts of enclosed embankments were stated explicitly to the respondents, they would not prefer

Table V. Respondents' assessments of positive and negative impacts of the CPP: inside vs outside pairwise comparisons (data are aggregates of Likert scale scores)

Items assessed	Inside Cluster 1b		Outside Cluster 1b		
Village	1	2	3	4	
				Pairwise difference	
Effectiveness of the project for flood alleviation				N = 200*	$p < 0.0001$
Mean	3.7	3.6	3.5	2.8	1, 2 and 3
Std dev	0.9	0.9	0.8	0.7	from 4
Worsening of flood outside the project				N = 200**	$p < 0.0001$
Mean	2.2	2.1	4.0	3.1	1 and 2 less
Median	2.0	1.0	4.0	4.0	than 3 and 4
Reduction of floodplain fishing				N = 200*	$p < 0.0001$
Mean	2.8	3.2	2.5	3.5	4 from 1
Std dev	1.0	1.1	0.9	0.8	
Concern about drainage congestion				N = 200**	$p < 0.3$
Mean	1.9	1.9	1.9	1.6	
Median	1.0	1.0	1.0	1.0	

*ANOVA tests (Tukey's Honestly Significant Difference pairwise t-tests).
**Kruskal-Wallis H tests (Bonferronni adjusted U-tests for pairwise comparisons).

engineering intervention by the compartmentalization scheme. The MDC experiments were designed to test these hypotheses, and to measure the importance of any one of these flood management outcomes when they were evaluated in the context of each other.

Table VI presents the results of the MDC experiment. The table reports the results for the inside communities only, because no significant differences were observed between communities located inside or outside of the CPP. See Figure 5 for a graphic inside vs. outside comparison of part-worth utilities. The analysis is based on a conditional multinomial logit regression conducted with NteLogit software (Intelligent Marketing Systems, 1991–93). The table lists parameter estimates, standard errors, t-values and significance levels for each attribute level. In MDC experiments, the relative importance of each variable level is assessed by treating the aggregate frequencies of the number of times each variable level is selected as the 'most' and the 'least' important as the dependent variable, while the attribute levels serve as independent variables (Louviere *et al.*, 1993). One attribute level from the entire list is selected as the base for the analysis (which in this case was the incidence of flooding inside homes and on the courtyards), and therefore it is constrained to zero. The parameter estimates reflect the relative weight of each attribute level in relation to the base, and are therefore directly comparable over the entire model. With a McFadden's Rho Square of 0.65, the model shows a very

Table VI. Results of the MDC Experiments: Multinomial logit model for inside Cluster 1b

Attribute /Attribute levels	Parameter estimate	SE of parameter	t-value	*p*
HOMESTEAD:				
Neither house nor yard flooded	8.0424	0.3306	24.322	0.0001
Yard flooded	2.8294	0.2053	13.778	0.0001
House and yard flooded	0.0000	→constrained to this value!		
WATER LEVEL IN AMAN RICE FIELD:				
2 ft	10.4544	0.3613	28.932	0.0001
5 ft	2.4559	0.1808	13.578	0.0001
8 ft	−0.2391	0.1758	−1.360	0.1740
EFFECTS ON FISH:				
No change in floodplain fish or in culture fish	5.9845	0.3082	19.415	0.0001
Increase in culture fish	4.2260	0.2988	14.140	0.0001
Decrease in floodplain fish and in culture fish	3.6176	0.2072	17.4555	0.0001
DRAINAGE:				
No problem with drainage congestion	7.5449	0.3112	24.238	0.0001
Drainage congestion due to malfunctioning of sluices	2.0680	0.1719	12.024	0.0001

Rho-square = 0.65.

good fit. (McFadden's Rho Square is a 'goodness of fit measure' for multinomial logit model, similar to R^2 in multiple regression).

An evaluation of all parameter estimates indicates that within each variable the estimates for the respective attribute levels make intuitive sense. As expected, respondents preferred no flooding of their homestead compared to the yard being flooded, and really disliked flooding of both the yard and the home. In the *aman* rice fields, the model produced a strong preference for the optimal flooding depth of 2 feet (0.6 m), while the two other depths were considered as less desirable. For the effects on fish, differences between the three levels were less pronounced, but no change was preferred over an increase in culture fish, and a decrease in both culture fish and floodplain fish was not considered any worse. Finally, as expected, a situation without drainage problems was significantly preferred over one with malfunctioning of sluices.

The interpretation of these results becomes more insightful when these estimates are compared across all variables. It now becomes obvious that within the entire model, the two estimates of no flooding on the homestead and 2 feet of

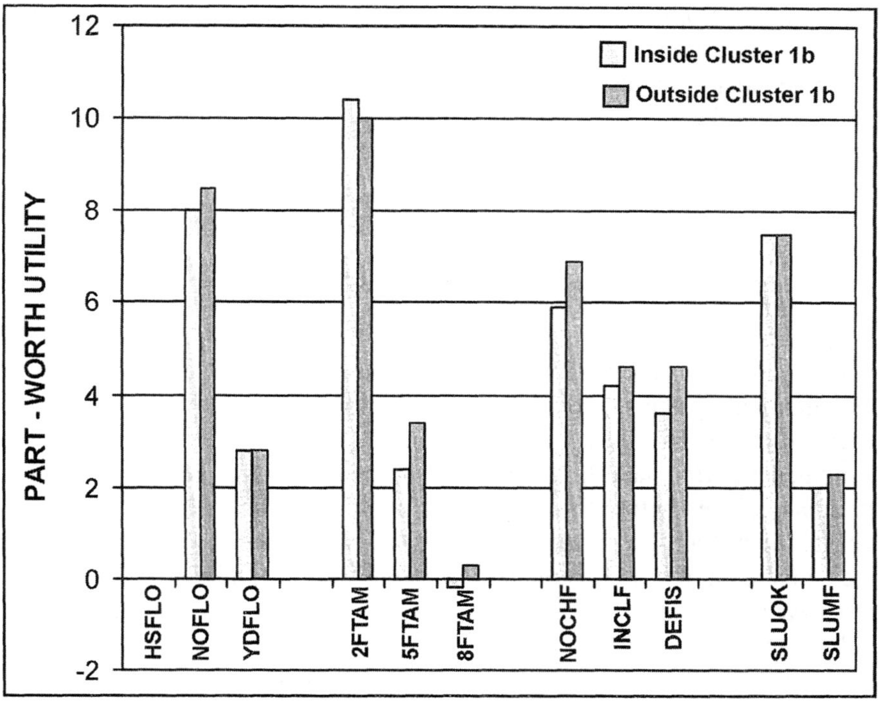

Figure 5. Part-worth utility of respondents' preferences for the outcomes of water level management by location (inside vs. outside Cluster 1b).

flooding in the *aman* field were by far the two most desirable variables. In fact, the ideal water level in the *aman* field was considered slightly more desirable than a flood-free homestead. On the other end of the spectrum were the situations of the home and the yard being flooded and a flood of 8 feet in the *aman* field, both of which were by far the least attractive outcomes. Of specific interest are the estimates for the remaining attribute levels in between these extremes. Respondents to the MDC considered no change in floodplain or culture fish and no problems with drainage congestion as highly desirable, as these estimates are positioned clearly between the two most desirable outcomes and all of the other levels. Declines in the fishery, as represented by the two other fishing levels, were perceived rather indifferently. However, malfunctioning of sluice gates was considered as slightly worse than a situation in which the yard would be flooded or the *aman* fields would be flooded to a depth of 5 feet of water. These latter observations are important, because obviously respondents were concerned about the malfunctioning of sluices and to a lesser extent about the declining fisheries when they were asked to evaluate management outcomes in full profiles rather than each variable separately. It should be stressed here that this finding might seem to contradict a relatively low concern about drainage congestion in the CPP area (reported earlier in Table V as mean Likert scale scores of 1.6–1.9). The latter might be the result of respondents' observations of no major problems with the current drainage conditions in the CPP area. In contrast, in the joint evaluation of several variables in the MDC experiment, the respondents made a subtle but powerful statement that drainage conditions do matter: third highest part-worth utility (7.5) for "no problems with drainage congestion". They would certainly prefer regulated flooding but not at the expense of drainage congestion.

6. Management Implications and Conclusion

It appears from the preceding findings of the MDC experiment that the floodplain residents of the CPP area would prefer regulated flooding inside enclosed embankment systems, provided there were no major drainage congestion inside such structures. However, based on earlier experiences with polders and other embankments in Bangladesh, the prospect for achieving this goal does not seem to be promising. Even before the completion of all engineering interventions in Cluster 1b, its main drainage regulator had been causing some backup of water immediately upstream of it. In several earlier studies, serious drainage problems have been reported from numerous polders in different parts of Bangladesh (Adnan, 1991; Adnan and Sufiyan, 1993). The problem with drainage of standing floodwater has also been reported along the Brahmaputra Right Bank Embankment during the 1988 flood (NEDECO, 1993, p. 51). Another study on the DND project, situated at the outskirts of the Dhaka Metropolitan Area, has shown that stagnant water inside this freshwater polder has induced a number of environmental problems, such as poor sanitation, water-borne diseases and an increase in mosquitoes (Rasid and Mallik, 1993; Rasid, 1996). A more recent study on the construction-phase drainage congestion along the Dhaka City Flood Control Embankment (Flood Action Plan 8: Dhaka Town Protection), perhaps, provided the most convincing evidence of potential drainage problems inside any enclosed embankment systems (Rasid and Mallik, 1996).

The proponents of the compartmentalization scheme have argued for increasing culture fish in ponds and dug pits to compensate for the loss of open-water floodplain fish (Brammer, 1990). The results of this study have largely contradicted this assumption. The residents of the CPP area were not willing to tolerate too much change in floodplain fish. Their preference for an increase in culture fish was significantly lower than their willingness to accept changes in floodplain fish.

While much of the social science research on floodplain residents in Bangladesh has so far dealt with perceptions of floods and preferences for structural and non-structural flood alleviation measures, this research included the environmental effects of engineering interventions. The present research was inspired by the findings of some of the earlier studies, which indicated that the floodplain residents of Bangladesh had an overwhelming preference for regulated flooding. However, it was not entirely clear if they preferred regulated flooding so badly that they were willing to accept its detrimental environmental effects, or if they would entertain tradeoffs between the benefits of structural flood control measures and environmental degradation associated with these measures. In order to research this question, this paper differed from previous research in two important ways. Firstly, rather than making the actual technical approaches as the focus of a survey with lay people, the study focused on outcomes of flood control measures. Secondly, instead of applying one of the traditional social science methods, such as researching preferences with Likert scales, this study applied a multivariate research approach focusing on stated preferences, in which respondents evaluated outcome scenarios in full profiles.

The successful application of a stated choice survey technique that elicits responses to hypothetical outcome scenarios in a multivariate manner has important implications for management and future social research in developing countries. First of all, the fact that respondents were able to complete such relatively complex tradeoff tasks successfully is important for potential future applications in developing countries. Presenting hypothetical scenarios in pictograms, perhaps, made the issues comprehensible to the respondents, many of whom had very little formal education. Secondly, such a structured and formal method should constitute an important complementary component of the more qualitative research, which is typical of all rural appraisal methods. The complementary nature between these two approaches is documented successfully in this research, as the insights generated by an RRA were used to define the variables and variable levels for the MDC.

The success of the application of the model can further be explained in part through reference to the ongoing testing and monitoring programs in the CPP by the Government of Bangladesh and its consultants (Euroconsult *et al.*, 1995). In particular, the variables used in the MDC experiments were consistent with the consultants' experimental variables, namely agriculture, fishery, environmental impact, and drainage problems. Whereas the consultants' study used specific technical data, some of these data and concepts were transformed into categorical variables for the MDC model. That is why, most of the results of the MDC experiments are consistent with the consultants' experiments. For example, respondents' support for 0.6 m of water in the *aman* rice field equated approximately with the consultants' ongoing experiments with regulated water levels in rice fields in some of the experimental sub-compartments.

The model results could potentially be used in two complementary ways. First, if more participatory decision-making processes were undertaken, then these model results could be presented to the participants. Second, if decisions were predominantly made by experts, then these managers or experts could consult such a model for understanding the tradeoff behaviour of their constituency.

Based on the successful MDC experiments, the current authors further developed and adopted a discrete choice experiment (DCE) in a 2000–2001 follow-up survey for assessing floodplain residents' preferences for alternative flood alleviation measures, namely compartmentalization vs poldering. The profiles of the

DCE choice cards included impacts of regulated water levels on rice crops and homesteads, the amount of disaster relief, and some of the adverse environmental impacts of enclosed embankments. Several of the pictograms of the MDC choice cards were adopted for this DCE. The preliminary results of this survey among 360 respondents from the Ganges, the Brahmaputra and the Meghna floodplains indicated that nearly 50% preferred compartmentalization, 40% opted for poldering, and only 10% opposed embankments. Thus, the main finding of the DCE survey reinforced that of the MDC experiment in support of floodplain residents' preferences for regulated flood regimes. However, the CPP residents were not willing to tolerate some of the suggested environmental impacts of enclosed embankment systems, notably changes in floodplain fish and major problems with drainage congestion. Devising appropriate engineering interventions, that are capable of minimizing such adverse impacts, seems to be the main challenge for water resources engineers.

In the Western democratic tradition, public participation in decision-making processes is considered as a cornerstone of sustainable environmental management (Dearden and Mitchell, 1998; Kundzewicz, 1997 and 1999). The multi-attribute stated preference models seem to be an appropriate approach for obtaining public input in an objective and representative manner (Haider and Rasid, 1998). Although the present study has demonstrated that these models could be applied successfully even in the third world cultural setting, the findings of this study should be treated with caution. Certainly, this study has proven conclusively that there is widespread support among the floodplain residents of Bangladesh for regulated flood regimes through different types of embankments. However, if engineering interventions are not capable of addressing the environmental impacts of enclosed embankment systems in a satisfactory manner, then stakeholder support alone could not be the basis for adopting such a controversial megaproject. This reservation is partly based on one of the limitations of stated preference models, that is, conclusions drawn from such models are still conditional on the scenario presented (Blamey *et al.*, 1999, p. 353). An overwhelming support for enclosed embankments in Bangladesh raises two further questions regarding the nature of this investigation. Firstly, what would be the respondents' preferences if scenarios of adverse impacts of enclosed embankments were presented in greater details? So far, the adverse impacts of enclosed embankments included loss of floodplain fish and drainage congestion. Many other impacts were not included in the tested MDC and the DCE models, primarily because of the design limitations. Most of the adverse impacts presented in the DCE were adopted from the MDC model. Among some of the excluded impacts, extensive poldering in the southwestern region of Bangladesh has induced accelerated siltation outside poldered areas. This has hindered efficient drainage of water from the floodplains (Mirza, 2000). Secondly, another potential explanation of support for embankments is more philosophical in nature. It is possible that the farmers of the CPP area, like most of the floodplain residents in Bangladesh who preferred regulated flood levels, had an implicit faith in engineering intervention, presuming that it was capable of solving their flood problems (Rasid and Hufferd, 1989; Rasid and Mallik, 1995). As indicated earlier, taped conversations with farmers supported this assumption.

In 1998 Bangladesh was engulfed in the largest flood of the century, surpassing the two previous records of 1987 and 1988 floods. Over two-thirds of the country (including large parts of the CPP) was inundated for a duration of nearly three months (Ahmad *et al.*, 2000). Water resources management authorities in Bangladesh, particularly Bangladesh Water Development Board, are constantly in search of acceptable flood alleviation measures. Flood problems in Bangladesh pose a difficult challenge for engineers and planners. Perhaps a stopgap measure lies between "living with the flood" and limited experimental engineering interventions, such as the CPP.

Acknowledgments

Financial support for this study was provided by a grant from the Social Sciences and Humanities Research Council of Canada (SSHRC Grant Number: 410-96-0227). Khan Manjur Morshed conducted door-to-door interviews. The manuscript benefited from helpful comments from two referees. Len Hunt (Ontario Ministry of Natural Resources) assisted with the statistical tests. Special thanks are due to Dr Don Anderson (StatDesign) for his contribution in the design and analysis of the MDC. Cathy Chapin drafted all final diagrams.

References

Adnan, S.: 1991, *Floods, People and the Environment*, Research & Advisory Services, Dhaka, Bangladesh.

Adnan, S. and Sufiyan, A. M.: 1993, *State of the FAP: Contradictions Between Policy Objectives and Plan Implementation*, Research & Advisory Services, Dhaka, Bangladesh.

Ahmad, Q. K., Chowdhury, A. K. A., Imam, S. H., and Sarker, M. (eds.): 2000, *Perspectives on Flood 1998*, The University Press Ltd, Dhaka, Bangladesh.

Ben-Akiva, M. and Lerman, S.: 1985, *Discrete Choice Analysis: Theory and Application to Travel Demand*, MIT Press, Cambridge, MA.

Blamey, R., Gordon, J., and Chapman, R.: 1999, Choice modeling: Assessing the environmental values of water supply options, *The Australian Journal of Agricultural and Resource Economics* **43**, 337–357.

Boyce, J. K.: 1990, Birth of a megaproject: Political economy of flood control in Bangladesh, *Environmental Management* **14**, 419–428.

Brammer, H.: 1990, Flood in Bangladesh. II Flood mitigation and environmental aspects, *The Geographical Journal* **156** (Part 2), 158–165.

Burton, I., Kates, R. W., and White, G. F.: 1993, *The Environment as Hazard*, 2nd edn., Oxford University Press, New York.

Chambers, R.: 1994, The origins and practice of Participatory Rural Appraisal, *World Development* **22**(7), 953–969.

Chowdhury, J. U.: 1991, Flood Action Plan: One sided approach? *Bangladesh Environmental News* **2** (2), 1–3.

Cochran, W. G. and Cox, G. M.: 1968, *Experimental Designs*, 2nd edn., Wiley, New York.

Custers, P.: 1992, Banking on a flood-free future? Flood mismanagement in Bangladesh, *The Ecologist* **22**(5), 241–247.

Dearden, P. and Mitchell, B.: 1998, *Environmental Change and Challenge: A Canadian Perspective*, Oxford University Press, Toronto.

Dempster, J. I. M. and Brammer, H.: 1992, Flood Action Plan – Bangladesh, *Outlook on Agriculture* **21**, 301–305.

Euroconsult, Lehmeyer International, Bangladesh Engineering & Technological Services, House of Consultants: 1995, *Compartmentalization Pilot Project: Water Management Workshop, Experience of the 1995 Monsoon*, Technical Note 95/07 (December 1995), Government of the Peoples Republic of Bangladesh, Ministry of Water Resources, Bangladesh Water Development Board, Flood Plan Coordination Organization.

European Parliament: 1993, Development or disaster?: Taming two of the mightiest rivers of the world; Test case for global disaster management coordinated by the World Bank, Conference on Flood Action Plan in Bangladesh, held at European Parliament, 27–28 May 1993.

Fischoff, B., Lichtenstein, S., Slovic, P., Derby, S. L., and Keeney, R. L.: 1981, *Acceptable Risk*, Cambridge University Press, New York.

Haider, W. and Hunt, L.: 1997, Remote tourism in northern Ontario: Patterns of supply and a motivational segmentation of clients, *Journal of Applied Recreation Research* **22**(1), 49–78.

Haider, W. and Rasid, H.: 1998, Assessing lay preferences for water level management in a regulated river system in northwestern Ontario: Application of a stated preference model, *Canadian Water Resources Journal* **23**(3), 83–101.

Haque, C. E. and Zaman, M. Q.: 1993, Human response to riverine hazards in Bangladesh: A proposal for sustainable floodplain development, *World Development* **21**(1), 93–107.

Hughes, R., Adnan, S., and Dala-Clayton, B.: 1994, *Floodplains or Floodplans?: A Review of Approaches to Water Management in Bangladesh?* Institute of Environment and Development, London and Research and Advisory Services, Dhaka, Bangladesh.

Johnson, B. L. C.: 1982, *Bangladesh*, 2nd edn., Heinmann Educational Books, London.

Kates, R. W. and Burton, I.: 1986, *Themes From the Work of Gilbert F. White: Geography, Resources and Environment*, Volume 2, University of Chicago Press, Chicago, IL.

Kundzewicz, Z. W.: 1997, Water resources for sustainable development, *Hydrological Sciences Journal* **42**(4), 467–480.

Kundzewicz, Z. W.: 1999, Flood protection-sustainability issues, *Hydrological Sciences Journal* **44**(4), 559–571.

Louviere, J. J. and Woodworth, C. G.: 1983, Design and analysis of simulated consumer choice or allocation experiments: An appraisal based on aggregate data, *Journal of Marketing Research* **20**, 350–367.

Louviere, J. J., Anderson, D. A., and Edwards, D. L.: 1993, *Maximum Difference Conjoint Theory, Methods and Comparisons with Choice- and Rating-Based Conjoint*, Working Paper, Department of Marketing, University of Utah, Salt Lake City, UT.

McFadden, D.: 1974, Conditional logit analysis of qualitative choice behaviour, in P. Zarembka (ed.), *Frontiers of Econometrics*, Academic Press, New York, pp. 105–142.

Mirza, M. M. Q.: 2000, The recent floods: What went wrong? *The Daily Star* (Bangladesh English daily) **3** (409), Thursday, October 19, 2000.

NEDECO: 1993, Flood Action Plan, Bangladesh: Evaluation Report 1993, Netherlands Development Cooperation, The Hague.

Paul, B. K.: 1984, Perception of and agricultural adjustments to floods in Jamuna floodplain, Bangladesh, *Human Ecology* **12**, 3–19.

Paul, B. K.: 1995, Farmers' response to the flood action plan (FAP) of Bangladesh: An empirical study, *World Development* **23**(2), 299–309.

Palm, R. I.: 1990, *Natural Hazards: An Integrated Framework for Research and Planning*, The Johns Hopkins University Press, Baltimore, MD.

Rasid, H.: 1996, Impact assessments from survey of floodplain residents: The case of the Dhaka-Narayanganj-Demra (DND) Polder, Bangladesh, *Impact Assessment* **14**(2), 115-132.

Rasid, H.: 2000, Reducing vulnerability to flood disasters in Bangladesh: Compatibility of floodplain residents' preferences for flood alleviation measures with indigenous adjustments to floods, in D. J. Parker (ed.), *Floods: Volume II* (Routledge Hazards and Disasters Series), Routledge, London, pp. 46–65.

Rasid, H. and Paul, B. K.: 1987, Flood problems in Bangladesh: Is there an indigenous solution? *Environmental Management* **11**, 155–173.

Rasid, H. and Hufferd, J.: 1989, Hazards of living on the edge of water: the case of Minnesota Point, Duluth, *Human Ecology* **17**, 85–100.

Rasid, H. and Mallik, A.: 1993, Poldering vs compartmentalization : The choice of flood control techniques in Bangladesh, *Environmental Management* **17**, 59–71.

Rasid, H. and Mallik, A.: 1995, Flood adaptations in Bangladesh: is the compartmentalization technique compatible with indigenous adjustments of rice cropping to flood regimes? *Applied Geography* **15**(1), 3–7.

Rasid, H. and Mallik, A. U.: 1996, Living on the edge of stagnant water: An assessment of environmental impacts of the construction-phase drainage congestion along the Dhaka City Flood Control Embankment, Bangladesh, Environmental Management **20**(1), 89-98.

Slovic, P., Kunreuther, H., and White, G.F.: 1974, Decision processes, rationality, and adjustments to natural hazards, in G. F. White (ed.), *Natural Hazards: Local, National, Global*, Oxford University Press, London, pp. 187–205.

Smith, K.: 1992, *Environmental Hazards: Assessing the Risk and Reducing Disaster*, Routledge, London.

Timmermanns, H.: 1984, Decompositional multivariate preference models in spatial choice analysis: A review of some recent developments, *Progress in Human Geography* **8**, 189–221.

UNDP/GOB (United Nations Development Program/Government of Bangladesh): 1989, *Bangladesh Flood Policy Study: Final Report*, UNDP and Government of Bangladesh Ministry of Irrigation and Flood Control, Dhaka, Bangladesh.

White, G. F.: 1945, *Human Adjustment to Floods: A Geographical Approach to the Flood Problem in the United States*, Research Paper No. 29, Department of Geography, University of Chicago, Chicago, IL.

White, G. F. (ed.): 1974, *Natural Hazards: Local, National, Global*, Oxford University Press, London.

Whyte, A. V. T.: 1986, From hazard perception to human ecology, in R. W. Kates and I. Burton (eds.), *Themes from the Work of Gilbert F. White: Geography, Resources and Environment*, Volume 2, University of Chicago Press, Chicagi, IL, pp. 240–271.

World Bank: 1990, *Flood Control in Bangladesh: A Plan for Action*, Technical Paper Number 119, Asia Region Technical Department, The World Bank, Washington, DC.

Natural Hazards **28:** 131–143, 2003.

Flood Management in India

P. K. MOHAPATRA[1] and R. D. SINGH[2]
[1]*Department of Civil and Environmental Engineering, University of South Carolina, Columbia, USA, E-mail: pkm@cc.nih.ernet.in;* [2]*National Institute of Hydrology, Roorkee, India, 247 667*

(Received: 30 August 2001; accepted: 4 March 2002)

Abstract. In this paper, flood problems in India, regional variability of the problem, present status of the ongoing management measures, their effectiveness and future needs in flood management are covered. Flood problems in India are presented by four zones of flooding, viz. (a) Brahmaputra River Basin, (b) Ganga River Basin, (c) North-West Rivers Basin, and (d) Central India and Deccan Rivers Basin. Some special problems, related to floods like dam break flow, and water logging in Tal areas, are also mentioned. Progress of various flood management measures, both structural and non-structural, are discussed. In addition, future needs to achieve efficient and successful flood management measures in India are also pointed out.

Key words: Flood management, flood problems, flooding zones, India, special floods

1. Introduction

Among all natural disasters, floods are the most frequent to be faced by India. Floods in the eastern part of India (Orissa, Bengal, Andhra Pradesh, Bihar) in the recent past, are striking examples. The annual precipitation, in India, including snowfall is estimated at 4,000 Billion Cubic Meter (BCM). Out of this, the seasonal rainfall in monsoon is of the order of 3,000 BCM. The rainfall in India shows great temporal and spatial variations, unequal seasonal distribution and geographical distribution and frequent departures from the normal. As per the information published by different government agencies, flood prone area in India has been increasing dramatically. As reported by Central Water Commission (CWC) under Ministry of Water Resources, government of India, the annual average area affected by floods is 7.563 million ha. This observation was based on data for the period 1953 to 2000, with variability ranging from 1.46 million ha in 1965 to 17.5 million ha in 1978. On an average, floods have affected about 33 million persons between 1953–2000. There is every possibility that this figure may increase due to population growth. Flood data for 1953–2000 published by CWC (IWRS, 2001) is presented in Appendix 1. The National Flood Commission (1980) has reported that the total flood prone area of India was 34 million ha. It has also mentioned that an area of 10 million ha has been protected, but the effective protection may be available to only 6 million ha. The area liable to flood is shown in Figure 1.

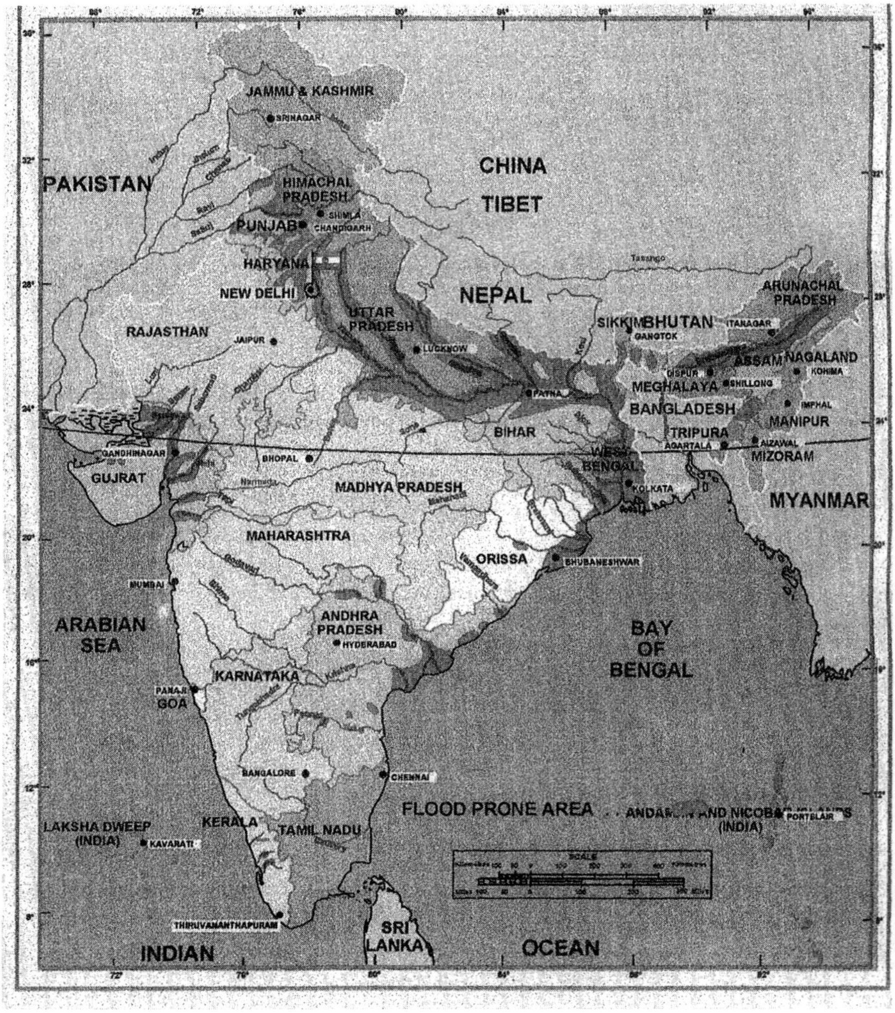

Figure 1. Flood prone areas in India.

Major causes of floods in India include inadequate capacity within riverbanks to contain high flows, riverbank erosion and silting of riverbeds. In addition, other factors are; land slides leading to obstruction of flow and change in the river course, retardation of flow due to tidal and backwater effects, poor natural drainage in the flood prone area, cyclone and associated heavy rainstorm/cloud bursts, snowmelt and glacial outbursts, and dam break flow.

Main objectives of this paper are to present (1) Flood problems in India, (2) Progress of on-going flood management measures, and, (3) Future needs in flood management.

2. Flood Problems in India

Main problems in India with respect to floods are inundation, drainage congestion due to urbanization and bank erosion. The problems depend on the river system, topography of the place and flow phenomenon. Being a vast country, the flood problems in India may be visualized on regional basis. However, for the sake of simplicity, India may be broadly divided into four zones of flooding, viz. (a) Brahmaputra River Basin, (b) Ganga River Basin, (c) North-West Rivers Basin, and (d) Central India and Deccan Rivers Basin. Flooding in these zones is presented in following subsections.

2.1. BRAHMAPUTRA RIVER BASIN

The first zone belongs to the basins of the rivers Brahmaputra and Barak with their tributaries. It covers the States of Assam, Arunachal Pradesh, Meghalaya, Mizoram, northern parts of West Bengal, Mainpur, Sikkim, Tripura and Nagaland. The catchments of these rivers receive large amount of rainfall. As a result of this, floods in this region take place very often and are severe by nature. The general tectonic up wrapping of North-East region has also significant effect on river Brahmaputra. Almost all Northern tributaries of Brahmaputra are affected by landslides in the upper catchment. Further, the rocks in the hills, where these rivers originate are friable and susceptible to erosion and thereby cause exceptionally high silt charge in the rivers. In addition, the region is subject to severe and frequent earthquakes causing numerous landslides in the hills, which upsets the regime of the rivers. Important problems in this region are flood inundation due to spilling of banks, drainage congestion due to natural as well as man-made structures and change of river flow. In recent years, the erosion along the banks of the Brahmaputra was enormous and has become a serious concern among the water resources engineers.

Main problems of flooding in Assam are inundation caused by spilling of the rivers Brahmputra and Barak as well as their tributaries. In addition, the erosion along the Brahmaputra is a serious problem. In Northern parts of West Bengal, the rivers Teesta, Torsa and Jaldakha are in floods every year and inundate large areas. During flooding, these rivers carry large amount of silt and have a tendency to change their courses. The rivers in Manipur spill over their banks frequently. The lakes in the territory are filled up during the monsoon and spread to large marginal areas. In Tripura, flood problems are the spilling and erosion by rivers.

2.2. GANGA RIVER BASIN

The Ganga and its many tributaries (the Yamuna, the Sone, the Ghaghra, the Gandak, the Kosi and the Mahananda) constitute the second zone. This zone covers Uttaranchal, Uttar Pradesh, Bihar, south and central parts of West Bengal, parts of Haryana, Himachal Pradesh, Rajasthan, Madhya Pradesh and Delhi. The normal annual rainfall of this region varies from about 60 cm to 190 cm of which more than 80 per cent occurs during the South-West monsoon. The rainfall increases from west to east and from south to north.

The flood problem is mostly confined to the areas on the northern bank of the Ganga River. The damage is caused by the northern tributaries of the Ganga by spilling over their banks and changing their courses. Though the Ganga is carrying huge discharges (57,000 to 85,000 m^3/s), the inundation and erosion problems are confined to some specific places only. In general, the flood problem increases from west to east and from south to north. In the north-western parts of the region, there is the problem of drainage congestion. The drainage problem also exists in the southern parts of West Bengal. The problem becomes acute when the main river, in which the water is to be drained, already has high water level. The flooding and erosion problem is serious in Uttar Pradesh, Bihar and West Bengal. In Rajasthan and Madhya Pradesh, the problem is not so serious. In Bihar, the floods are largely confined to the rivers of North Bihar and are an annual feature. Most of the rivers (e.g., the Burhi Gandak, the Bagmati, the Kamla Balan, other smaller rivers of the Adhwra Group, the Kosi in the lower reaches and the Mahananda at the eastern end) spill over their banks causing considerable damage to crops and dislocating traffic. High floods occur in the Ganga occasionally causing considerable inundation of the marginal areas in Bihar.

In Uttar Pradesh and Uttaranchal, the flooding is frequent in the eastern districts, mainly due to spilling of the Rapti, the Sarada, the Ghaghra and the Gandak. The problem of drainage congestion exists in the western and north-western areas of Uttar Pradesh, particularly in Agra, Mathura and Meerut districts. Erosion is experienced in some places on the left bank of Ganga, on the right bank of the Ghaghra and on the right bank of the Gandak. In Haryana, flooding takes place in the marginal areas along the Yamuna and the problem of poor drainage exists in some of the south western districts. In Delhi, a small area along the banks of the Yamuna is subject to flooding by river spills. In addition local drainage congestion is experienced in some of the developing colonies during heavy rains. In the south and central part of West Bengal, the Mahananda, the Bhagirathi, the Ajoy, and the Damodar cause flooding due to inadequacy in river channels and tidal effect. There is also the problem of erosion of the banks of rivers and on the left and right banks of Ganga, upstream and downstream, respectively, of the Farakka barrage.

2.3. NORTH-WEST RIVER BASINS

This is the third zone and comprises of basins of North-West rivers such as the Sutlej, Ravi, Beas, Jhelum and Ghaggar. In comparison to the two zones mentioned above, the flood problem in this zone is relatively less. The major problem is that of inadequate surface drainage which causes inundation and water logging.

Another cause for flood has been the water logging in the irrigated area and changes in river regimes due to increased ground water levels. At present, the problems in Haryana and Punjab are mostly of drainage congestion and water logging. The Ghaggar River used to disappear in the sand dunes of Rajasthan after flowing through Punjab and Haryana. The Jhelum floods occur frequently in Kashmir causing a rise in the level of the Wullar Lake thereby submerging marginal areas of the lake.

2.4. CENTRAL INDIA AND DECCAN RIVERS BASIN

Important rivers in the fourth zone are the Narmada, the Tapi, the Mahanadi, the Godavari, the Krishna and the Cauvery. These rivers have mostly well defined stable courses. They have adequate capacity within the natural banks to carry the flood discharge except in their lower reaches and in the delta area, where the average bed slope is very flat. The lower reaches of the important rivers on the East Coast have been embanked.

This region covers all the southern States namely Andhra Pradesh, Chhattisgarh, Karnataka, Tamil Nadu, Kerala, Orissa, Maharashtra, Gujarat and parts of Madhya Pradesh. The region does not have very serious problems except for some of the rivers of Orissa (the Brahmani, the Baitarni, and the Subarnarekha). The Delta areas of the Mahanadi, Godavari and the Krishna rivers on the east cost periodically face flood and drainage problems, in the wake of cyclonic storms.

The Tapi and the Narmada are occasionally in high floods affecting areas in the lower reaches of Gujarat. Flood problems in Andhra Pradesh is confined to spilling by the smaller rivers and the submergence of marginal areas along the Kolleru Lake. Rivers like Budameru and the Thammileru not only overflow their banks along their courses to Kolleru Lake but also cause a rise in the level resulting in inundation of adjoining lands.

In Orissa, damage due to floods is caused by the Mahanadi, the Brahmani and the Baitarani which have a common delta. Water from these rivers intermingles in the delta and results in a very high water level, which cause severe flooding in the region. The coastal districts are densely populated and receive heavy precipitation in the Eastern Ghat region. The silt deposited constantly by these rivers in the delta area raises the flood water level and, the rivers often overflow their banks or break through new channels causing heavy damage. The lower reaches of the Subarnarekha are effected by floods and drainage congestion. The small rivers of Kerala when in high floods often cause considerable damage. In addition, there is also the problem of mud-flow from the hills, which results in severe losses.

3. Special Problems

In this section, some special flood problems in India are presented.

3.1. PROBLEM OF TAL AREAS

Natural depressions where water gets deposited during monsoon for longer period are known as Tal areas. Generally, they hamper normal activity affecting the Kharif crop. Mokama group of Tal in Bihar is known for its flood problem. Water gets accumulated in these areas during monsoon and remains stagnant up to September. Similar problems are also seen in Ghaggar detension basin in Rajasthan, depressions available at Ottu, Bhindawas, Kotla lakes in Haryana. Flood problems in such areas are of special nature and needs to be treated separately.

3.2. RIVER BANK/BED EROSION

All natural rivers are with mobile bed flows. Therefore, depending on the flow phenomenon, there may be aggradation and/or degradation in the river banks and beds. A river erodes its banks due to various reasons causing considerable loss of land, deterioration of the river regime and sometimes account for huge losses during floods. Rivers in Brahmaputra-Barak and Ganga basins are prone to severe erosion. The erosion is governed by the discharge in the river, bed slope, sediment flow and composition of bed and bank materials. Deforestation of upper catchment and hills lead to increased sediment load in rivers. Effect of seismic disturbances, and topographical conditions of land also contributes to the erosion problem. River erosion causes a loss to the land resources. The river behaviour causes new riverine landmass to be built up, but these become productive after many years and cannot compensate the land-loss due to erosion. Erosion in the Majuli island, the largest river-island in the world, is the most appropriate example to state the severity of the problem. The Brahmaputra Board has estimated that in the Majuli Island, the annual loss of land due to erosion could be about 3.9 km^2 and an economic loss of about Rs. 31.5 million per annum (IWRS, 2001).

3.3. SEDIMENT TRANSPORT BY RIVERS

One of the problems associated with the floods in India is the transport of sediments by rivers during floods. Himalayan rivers originating from Nepal bring a lot of sediment during floods to the alluvial plains in the valley. Transport of sediments (suspended and bed load) has a major role on river behaviour and river morphology. Thus, the flood problem and its management measures depend a lot on sediment transport. Several observation-sites for sediment transport in rivers are maintained by the CWC. With the help of the data (on average suspended sediment load), the pattern of increase/decrease of silt load could be examined to assess the morphological changing trend of the river. As recorded by CWC, the total live and gross storage capacity created in India is about 177 and 217 km^3, respectively. Based on the sedimentation data of 144 reservoirs, the weighted average annual loss in gross storage due to silting computed is 0.44%. Thus likely annual loss in the total gross storage of 217 km^3 is 0.95 km^3. Similarly, the annual loss in live storage is 0.31%

based on the data of 42 reservoirs. Thus the likely annual loss in the total live storage is 0.55 km^3. Considering the average density of 1.137 tonnes/m^3, based on the data of 13 reservoirs, the weight of the total sediment deposits in all the reservoirs in India is 1,080 million ton annually.

3.4. DAM BREAK FLOWS

Flooding due to dam break is a mega-disaster as it is associated with huge loss of life and property. An unusual high peak in a short duration and presence of a moving hydraulic shock/bore make it a different problem as compared to other natural floods. In India, historical events (e.g., failure of Machhu dam, Panshet dam, Nanak Saagar dam) for dam break floods are common (Palaniappan, 1997). Sometimes, blockage of water due to deposits caused by landslide takes place. When this natural blockage fails due to increased amount of water at the upstream, huge flooding occurs. The behavior of this flood is similar to that of dam break floods.

3.5. URBAN DRAINAGE

Flooding of cities in India is a common and annual event. Due to encroachment of the flood plain areas, presence of several structures and absence of proper regulations for maintenance, artificial flood is created. Therefore, proper drainage networking for a city has to be developed.

3.6. FLASH FOODS

Flash floods are characterized by sudden rise and recession of flow of small volume and high discharge which causes damages because of suddenness. They generally take place in hilly region where the bed slope is very steep. Typical examples are flash flood of Arunanchal Pradesh and flash flood of Satluj in 2000. Large reservoir downstream of flood prone areas can absorb the flood wave.

3.7. FLOOD DUE TO SNOWMELT

Snowmelt is a gradual process. Usually, it does not produce floods. However, sometimes, glaciers hold large quantity of bounded water. When released suddenly, this causes severe flooding. The rivers originating from the Himalayas in the north are fed by snowmelt from glaciers. In 1929, the outburst of the Chong Khundam glacier (Karakoram) caused a flood peak of over 22,000 m^3/s at Attock.

3.8. FLOOD IN COASTAL AREAS

Floods in Indian river basins are also caused by cyclones. Coastal areas of Andhra Pradesh, Orissa, Tamilnadu, and West Bengal experience heavy floods regularly. The flood due to the super cyclone combined with heavy rainfall during October 1999 in the coastal region of Orissa is an example. During past 110 years (1891–2000), over 1,000 tropical cyclones and depressions, originating in the Bay of Bengal and Arabian Sea, moved across India. Passage of such storms over a river basin leads to severe floods.

4. Flood Management in India

4.1. POLICIES ON FLOOD MANAGEMENT

After the unexpected heavy floods in 1954, the Government of India took several steps to constitute a number of committees to study flood problems in India. Some of the important ones are;

1. Policy statement (1954).
2. High level committee on flood (1957).
3. Policy statement (1958).
4. Ministerial committee on flood control (1964).
5. Ministers committee on floods and flood relief (1972).
6. Working groups on flood control for five year plans.
7. Rashtriya Barh Ayog (1980).
8. National water policy (1987).
9. National commission for integrated water resource development plan (1996).
10. Regional task forces (1996).

The above mentioned commissions on flood have given valuable recommendations on different aspects of flood management. Many of the recommendations are applied in the field in India. However, some of these are not accepted by various agencies. In addition, it is also observed that some of the recommendations are not effective in Indian context.

4.2. CURRENT STATUS OF FLOOD MANAGEMENT IN INDIA

Progress of Structural Measures: After the disastrous floods in 1954, a National Programme of Flood Management was launched. During various plan periods, different methods of flood protection, both long-term and short-term, have been adopted in different states. The total length of embankment structures is 16,800 km and drainage channels are of 32,500 km. A total of 1,040 towns and 4,760 villages are currently protected against flood.

Major embankment projects taken up and completed after independence are on the following rivers: Kosi and Gandak (Bihar), Krishna and Godavari (Andhra Pradesh), Mahanadi (Orissa), Tapi (Gujarat), Lower Damodar (West Bengal),

Brahmaputra (Assam). These embankments have played important role in providing reasonable protection to affected areas. Barring occasional breaches in embankments, the works have given reasonable protection to an area of about 15.07 M ha. In addition, an area of about 3 M ha, which were already existing in some states, had been protected prior to 1954 by works, which were already existing in some of the states. A large number of reservoirs were constructed in India during various five-year-plan periods. These reservoirs have helped greatly in reducing the intensity of flooding downstream of the flood plains.

Progress of Non-Structural Measures: The response of the State Governments towards enactment of flood plain zoning bill is not encouraging. Though some of the states (e.g., Manipur and Rajasthan) have enacted the flood plain zoning legislation, major flood affected states (e.g., Assam, Goa, Himachal Pradesh, or Sikkim) have not considered such legislation. One of the main requirements for implementing the flood plain zoning as an effective flood management measure is availability of survey maps on suitable large scale to enable proper zoning of flooded areas. An area of about 55,000 km^2 of flood plains was surveyed. These surveys cover areas along the Ganga, Yamuna, Ramganga, Roopnarayan, Jalangi and other flood prone rivers of West Bengal, tributaries of Brahmaputra like Burhi Dihing, Desang, Dikhoo, Sutlej and Ravi. In total, 570 maps (1,09,267 sheets) were prepared.

A working group of National Natural Resources Management System (NNRMS) standing committee on water resources for flood risk zoning of major flood prone rivers considering remote sensing input was constituted by Ministry of Water Resources in 1999. The working group recommended flood risk zoning using satellite based remote sensing with a view to give thrust towards implementation of flood plain zoning measures in India.

Flood proofing measures taken in the past in India consisted of raising of a few flood prone villages, above pre-determined flood level and connecting them to nearby roads or high lands. Under this programme, several villages in West Bengal, Uttar Pradesh and Assam were raised. The programme of raising of villages was subsequently discontinued, as it was observed that the method could not protect the nearby agricultural areas. Flood forecasting and flood warning in India was commenced in a small way in 1958, for the Yamuna River at Delhi. This has evolved to cover most of the flood prone interstate river basins in India. Central Water Commission is presently responsible for issuing flood forecasts at 157 stations.

Planning and execution of flood management measures are primarily the works of the States and Union Territories. The State governments undertake flood management measures depending on availability of funds and priority. The Central government makes available funds for flood in annual plans to the States through its Planning Commission. In addition, the Central government gives special assistance to the border States and North-Eastern States for taking up some special priority works. Plan wise expenditure on flood management measures up to 8th plan period (ending 1997) is Rs. 5032.13 crores. The outlay for the 9th plan is Rs. 2938.49

crores. CWC (1996) has reported the recommendations of the working group on flood management for the 9th five year plan covering the period 1997–2002.

Ministry of Agriculture, government of India has prepared a contingency action plan for all natural disasters including floods. The Relief Commissioner and Additional Secretary of the above ministry is the nodal officer at the center. He receives all information from states and other agencies from the affected areas. Most of the states have Relief Commissioners who are in charge of relief measures. There is a National Crisis Management Committee (NCMC) with cabinet secretary as chairman. There is also a Crisis Management Group (CMG) with Relief Commissioner as the chairman. When situation is to be handled by NCMC, it gives necessary direction to CMG.

5. Future Needs in Flood Management

5.1. FOCUSED APPROACH

Though there is a considerable progress in various flood management measures in India, the problems due to floods still persist. This may be attributed to inadequacy of planned investment. Flood management measures has to be more focused and targeted towards the decided goals within a stipulated time frame. A flood management measure needs the river basin as a whole to be considered as unit for integrated planning for optimum utilization of the resources.

5.2. LEGISLATION FOR FLOOD PLAIN ZONING

As discussed earlier, many flood prone states have not adopted the recommendations regarding flood plain zoning. A detailed status for each state may be found in the report of the working group on flood management (CWC, 1996). Therefore, it is necessary to evolve methods in consultation with local bodies such that the legislation for flood plain zoning will be adopted. A task committee for the purpose is essential.

5.3. INADEQUACY OF FLOOD CUSHION IN RESERVOIRS

It is well recognized that long-term solution of flood problems lies in creating appropriate flood storage in reservoirs. The total live storage capacity of completed projects in India is 174 km^3. However, provision of flood storage capacity is limited to a few reservoirs like Damodar Valley Corporation (1.87 km^3), Ukai reservoir (1.33 km^3) and some smaller projects. There is no flood storage provided in major reservoirs like Hirakud. Thus, the flood storage capacity as a percentage of live storage capacity is negligible. Therefore, in future, large flood space is required for a successful flood management program.

5.4. FLOOD INSURANCE

Though flood risk has been included in the list of items covered by General Insurance Companies in India, it is popular in urban areas only. The insurance companies have also not been able to arrive at different rates of insurance premiums for different areas. Therefore, there is a need to evaluate the risk associated with flooding of different areas and to modify the insurance schemes based on it.

5.5. FLOOD DATA CENTER

Success of any research activities related to flood studies needs accurate and reliable data. Various flood management measures also depend on correct information about the flood. Therefore, an exclusive data center for flood management is required. Networking of nodal data centers is also required. Advanced techniques like remote sensing may be used to procure various data.

5.6. RESEARCH AND DEVELOPMENT

The Science and Technology Advisory Committee of the Ministry of Water Resources identifies and provides necessary guidelines and directions for Research and Development activities in water resources sector. Indian National Committee on Irrigation and Drainage is presently entrusted with the work of coordinating R&D schemes in flood management sector. Advanced knowledge in science and technology like application of remote sensing data etc. (Rao *et al.*, 1983) has to be used in studies for flood management.

5.7. COMMUNITY PARTICIPATION

With floods being a common concern for all sectors of society, flood management needs active involvement and participation of all to fulfill its objective. Farmers, professional bodies, industries, women, voluntary organizations are to be aware of flood management. Voluntary organizations can act like an interface between the government and public. People's participation in preparedness, flood fighting and disaster response is required. Media like TV and newspapers play an important role in flood.

5.8. INTERNATIONAL COOPERATION

India is drained by a number of international rivers that originate beyond its borders and flow into India. India shares river systems with six neighboring countries: Viz. Nepal, Bhutan, China, Myanmar, Bangladesh and Pakistan. Bilateral cooperation for various flood management measures is essential for India and the concerned country. Government of India has already taken some initiatives in this regard. However, more active participation in the subject is required.

6. Conclusion

In this technical paper, flood management in India was presented. Flood problems in India were described dividing the country into four zones, viz. (a) Brahmaputra River Basin, (b) Ganga River Basin, (c) North-West Rivers Basin, and (d) Central India and Deccan Rivers Basin. Specific phenomena related to flood in the respective zones were pointed out indicating the regional variability of the problem. Past and ongoing measures for flood management in India were presented with an emphasis to performance and progress of various measures, both structural and non-structural. In addition, future needs to achieve efficient and successful schemes for flood management were pointed out. Some special flood-problems, like dam break flow and flood in the Tal areas, were also mentioned.

Appendix 1. Damage due to Floods during 1953 to 2000

Year	Area affected (Th. ha.)	Population affected (Th.)	Human lives lost	Total monetary damage (Rs. Cr.)
1953	2290	24280	37	52
1954	7490	12920	279	57
1195	9440	25270	865	103
1956	9240	14570	462	54
1957	4860	6760	352	23
1958	6260	10980	389	44
1959	5770	14520	619	86
1960	7530	8350	510	63
1961	6560	9260	1374	31
1962	6120	15460	348	95
1963	3490	10930	432	37
1964	4900	13780	690	67
1965	1460	3610	79	7
1966	4740	14400	180	88
1967	7120	20460	355	155
1968	7150	21170	3497	211
1969	6200	33220	1408	404
1970	8460	31830	1076	288
1971	13250	59740	994	632
1972	4100	26690	544	158
1973	11790	64080	1349	569
1974	6700	29450	387	569
1975	6170	31360	686	472
1976	11910	50460	1373	889
1977	11460	49430	11316	1202
1978	17500	70450	3396	1455
1979	3990	19520	3637	614
1980	11460	54120	1913	841
1981	6120	32490	1376	1197
1982	8870	56010	1573	1645

Year	Area affected (Th. ha.)	Population affected (Th.)	Human lives lost	Total monetary damage (Rs. Cr.)
1983	9020	61030	2378	2492
1984	10710	54550	1661	1906
1985	8380	59590	1804	4059
1986	8810	55500	1200	3749
1987	8890	48340	1835	2570
1988	16290	59550	4252	4630
1989	8060	34150	1718	2405
1990	9303	40259	1855	1709
1991	6357	33889	1187	1488
1992	2645	19256	1533	3345
1993	11439	30409	2864	3282
1994	4805	27548	2078	1795
1995	5245	35932	1814	3702
1996	8049	44729	1803	3006
1997	4569	29663	1402	2831
1998	9133	68718	2758	5846
1999	3978	25659	576	2108
2000	4940	40063	2345	1416

References

Central Water Commission: 1996, Report of the Working Group on Flood Management for the 9th five year plan (1997-2002), CWC, New Delhi.

IWRS: 2001, Theme Paper on "Management of floods and droughts", Indian Water Resources Society, Roorkee, India, pp. 4–12.

National Flood Commission: 1980, Report of the Rashtriya Barh Ayog, New Delhi.

Palaniappan, A. B.: 1997, Dams and their failures, chapter 8, Lecture notes on "Dam Break Flow Analysis", National Institute of Hydrology, Roorkee, pp. 126–171.

Rao, V. R., Lalitha, L., Rao, P. P. N., Gupta, A. K., and Vijayalakshmi, G., 1983, Disasters in India – can remote sensing do something, Indian Space Research Organization, Bangalore, India.

Natural Hazards **28:** 145–154, 2003.

Glacial Lake Outburst Floods in the Nepal Himalaya: A Manageable Hazard?

RICHARD KATTELMANN
Watershed Management Council, P.O. Box 1090, Mammoth Lakes, CA 93546, USA; E-mail: rick@watershed.org

(Received: 13 April 2001; accepted: 4 March 2002)

Abstract. Within the past fifteen years, glacial lake outburst floods have become an active topic of discussion within the development community focused on Nepal. Such floods endanger thousands of people, hundreds of villages, and basic infrastructure such as trails and bridges. The flood risk is also a major impediment to hydroelectric development in several river basins. Unlike most other mountain hazards in Nepal, reducing the possibility of outburst floods is technically feasible. The first attempt within Nepal to reduce the hazard of one lake by artificially lowering its water level was partially completed in June 2000. Completing this task and beginning work on other hazardous lakes will require difficult decisions about risk by downstream residents and substantial investment from the international aid community.

Key words: outburst floods, glacial lakes, glaciers, Himalaya, Nepal, hazard mitigation

1. Introduction

Catastrophic floods originating from the outburst of glacial lakes have recently become recognized as one of the primary natural hazards in Nepal. Although other natural disasters such as rainfall-floods, earthquakes, landslides, wildfires, and epidemics have claimed the lives of thousands of Nepalis in recent decades (Oi, 1993; Poudyal-Chhetri, 1999), glacial-lake outbursts are feared for the potential devastation from a single large event. Following a destructive flood in the Khumbu Himal of eastern Nepal in 1985, several studies and inventories of Himalayan glacial lakes have been conducted, and the hazard is becoming known to agency professionals, politicians, and the general public. The Nepali press has carried dozens of articles in recent years, and stories about glacial lakes occasionally appear in international publications ranging from the *New York Times* (Wilford, 1988) to *New Scientist* (Pearce, 1999). As awareness of glacial lake outburst floods grows, people at risk along with technical experts and potential donors need to decide whether to pursue strategies to actively reduce the hazard posed by these lakes.

This short note seeks to briefly describe the continuing threat of outburst floods, draw attention to opportunities to reduce the hazard, and provoke debate about investing in hazard reduction. The author's interest in the topic is based on observations made opportunistically during six visits to several of the glacial lakes of

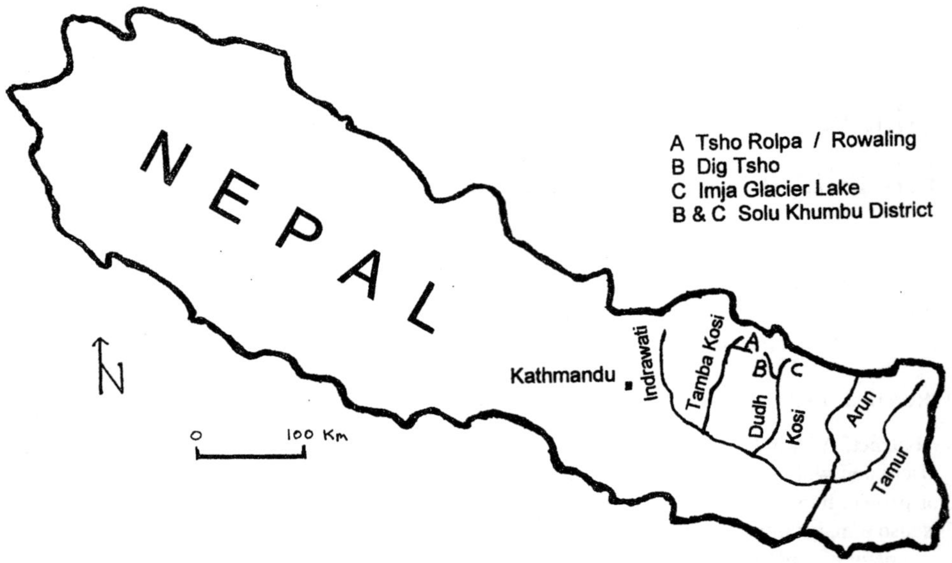

Figure 1. The rivers and sites of glacial lake outburst activity mentioned in the text are located in the eastern portion of Nepal.

eastern Nepal between 1982 and 1996. Some of the discussion relies on conversations with residents who observed the effects of outburst floods and who live with the threat of future floods. Only general impressions were used from these conversations.

2. The Nature and Hazard of Glacial Lake Outburst Floods

Glacial lakes are a common feature at altitudes of 4,500 to 5,500 m in many river basins of the Nepal Himalaya (Figure 1). Such lakes are found behind moraine dams, on the surface of glaciers, within glaciers, and where glacial ice and/or a lateral moraine blocks a side valley (e.g., Ives, 1986; Walder and Fountain, 1997). Moraine-dammed lakes tend to be the most common type in Nepal (Yamada, 1993 and 1998). As glaciers have rapidly retreated during this century (e.g., Rothlisberger and Geyh, 1985; Yamada, *et al.*, 1992; Evans and Clague, 1994), basins are created within the evacuated space between the retreating glacier and the lateral and end moraines. At the same time, melt-water ponds often form on the debris-covered snouts of these glaciers. These ponds may enlarge and coalesce into glacial lakes that lie on top of and adjacent to stagnant ice. As the glacier retreats and the lake grows, this so-called dead ice may both constitute part of the impoundment along with moraine material and contribute to the volume of the lake as the ice melts.

Some of the glacial lakes have been observed to grow at surprisingly rapid rates. The surface area of Tsho Rolpa (Figure 1) in the Rowaling Valley grew from about

0.23 km^2 in 1959 to 0.62 km^2 by 1972 and 1.3 km^2 by 1990 (Yamada, 1998). Imja Glacier lake increased from a few small ponds in the 1950s to a single body of water with a surface area greater than 0.5 km^2 in 1984 (Watanabe *et al.*, 1994). The ice-cored moraine and water surface elevation of Imja Glacier lake are also changing very rapidly as the ice melts (Watanabe *et al.*, 1995). The initial stages of rapid growth of a supraglacial lake on the Ngozumpa Glacier in the Khumbu Himal were observed in the late-1990s (Benn *et al.*, 2000).

Although moraine dams could be quite stable in a theoretical sense if the lateral and end moraines are relatively thick and lack imbedded ice, most should be viewed with suspicion from a hazard assessment viewpoint. The ice-cored moraines shift as the ice melts, are subject to internal piping, and can be incised rapidly by flowing water. Areas of ice that become exposed where the debris cover has slipped away absorb solar radiation and can melt rapidly, leading to structural degradation and mechanical failure (Richardson and Reynolds, 2000). Even moraine dams that have impounded small lakes for hundreds of years can eventually fail if they are overtopped or are subject to piping (Carling and Glaister, 1987).

Failure of these unstable moraine dams (and consequent release of large volumes of water within a few hours) occurs through two principal mechanisms. Waves generated by rockfall, snow and ice avalanches, glacier calving, or earthquakes can surge through the outlet channel and initiate rapid incision of the moraine. Alternatively, the dam may progressively weaken through seepage and enlargement of drainage conduits in the moraine until it is unable to hold back the impounded water. Moraine dams can also become weaker from seismic activity and removal of material by landslides. The greatest period of risk seems to be during the summer monsoon because of high water levels behind the moraines and high discharges over the moraine dams, resulting from rainfall and melt of ice and snow. Almost all the known outburst floods have occurred during the months of July, August, or September (Yamada, 1998).

At their extreme, glacial lake outburst floods can release millions of cubic meters of water in a few hours. Instantaneous discharges can be up to thousands of cubic meters per second. Obviously, such discharges have enormous power and the potential to do considerable geomorphic work. Large amounts of moraine material are initially entrained as the dam is breached. Although much of this material may be deposited in the first few hundred meters downstream, the flow is likely to become hyperconcentrated with particles from the moraine itself and the valley floor and side slopes. Landslides are initiated by removal of toe slopes and may temporary impound some of the flood flow. As these debris dams are breached, secondary outbursts occur, and a surging flow pattern may be established. Eventually, the flood waves are attenuated downstream, and material is deposited where gradients decrease. A critical distinction between outburst floods and rainfall-runoff or snowmelt floods is the point-source nature of the water in an outburst flood rather than the convergence of tributary contributions. Therefore, stream power and the capacity for erosional work decrease downstream in an outburst (Deslodges and

Church, 1992). After the initial effects of the flood itself, a valley may remain unstable for months to years as the channel and side slopes readjust.

3. Recent Glacial Lake Outburst Floods

Although glacial lake outbursts have obviously helped shape the Himalayan landscape over time, outburst floods in the eastern Himalaya were rarely mentioned in the western literature (e.g., Muller, 1959; Gansser, 1970) until the past two decades. In 1956, a large flood ascribed to a glacial lake outburst left 40,000 people homeless in the Indrawati valley, and a small glacier lake outburst flood occurred in the Imja valley (Muller, 1959). Occurrence of seven other outburst floods in rivers of eastern Nepal between 1935 and 1970 have been inferred from historical analyses (Yamada, 1998). An outburst flood in 1977 in the Khumbu Himal was recognized in the record of a stream gaging station 60 km downstream, and was the first event in the region to receive much scientific study. This flood originated from a series of lakes on and below the Nare Glacier in the Mingbo Valley west of the summit of Ama Dablam. Imagery from the LANDSAT Multi-Spectral Scanner before and after the event indicated that a small lake at about 5,160 m drained into a much larger proglacial lake and another smaller lake and caused the release of about 4.9 million m^3 of water (Buchroithner *et al.*, 1982). Another study of this flood estimated that only about 400,000 m^3 of water was released and that the initial failure was caused by melt of an ice-cored moraine (Fushimi *et al.*, 1985). Nevertheless, channel changes in the Imja Khola and Dudh Kosi were extensive, and parts of river terraces collapsed tens of kilometers downstream.

Another outburst event in the Khumbu Himal in August 1985 initiated much of the current concern about glacial lake outburst floods within Nepal as well as internationally because it destroyed a nearly-completed small hydroelectric project valued at about USD 4 million. This flood has been thoroughly researched and described in detail (Vuichard and Zimmermann 1986 and 1987; Ives, 1986). This flood occurred in the Bhote Kosi, western tributary to the upper Dudh Kosi (Figure 1). In the Langmoche Valley above the village of Thame, the lake Dig Tsho occupied the moraine-enclosed basin created by the retreat of the Langmoche glacier. It was about 1,500 m by 300 m and up to 18 m deep. The volume of water stored in the lake was about 5 million m^3 with another 1–2 million m^3 of dead ice (Vuichard and Zimmermann 1987). On August 4, 1985, a large ice avalanche impacted the lake and generated a wave about 5 m high that overtopped the moraine dam. The dam is believed to have eroded in stages with large boulders offering some resistance to instantaneous collapse (Vuichard and Zimmermann, 1987). Nevertheless, the lake drained almost completely in 4–6 hours with a peak discharge of about 2,000 $m^3\ s^{-1}$. Almost 900,000 m^3 of material was eroded from the moraine and mostly deposited within the first 2 km below the lake. Farther downstream, the surging flow eroded the river bed and banks in steeper sections and redeposited material in wider reaches where the slope decreased to 5 to 7 percent. Where large-volume collapses occurred, the flood was probably impounded or impeded until the landslide dam breached, resulting in intermittent surges down the channel. Channel surveys and mass balance estimates indicated that about 2.6 million m^3 of material were eroded and redeposited within the first 25 km and a total of about 3.3 million m^3 within first 40 km downstream of the lake (Vuichard and Zimmermann, 1987). Remarkably, only 4 or 5 people lost their lives in the flood itself because a Sherpa festival was in progress and few people were walking the trails at the time. A dozen bridges, more than two dozen homes, and several hectares of agricultural land were destroyed in addition to the hydroelectric facility. In the year following the Langmoche flood, debris flows and landslides mobilized only a small fraction of the material moved during the outburst but resulted in more deaths than during the original event (Watanabe *et al.*, 1994).

The Langmoche flood alerted the government of Nepal and the international agencies operating in Nepal to the hazard of glacial lake outburst floods. Analyses of hydrologic and geomorphic data and personal accounts revealed that three other outburst floods (in the Tamur, Sun Kosi, and Arun rivers) had occurred between the 1977 and 1985 events (Yamada, 1998). An outburst of Chubung lake in the Rowaling valley in July 1991 released about 0.5–1 million m^3 of water, eroding the river channel and damaging a few houses in the principal village of the valley (Damen, 1992). Three years later, an outburst occurred on the Pho Chhu of the Lunana area in the neighboring Bhutan. More than 20 deaths and significant property damage occurred 86 km downstream of the lake in the village of Punakha (Watanabe and Rothacher, 1996). Another outburst just south of the Khumbu region occurred on September 3, 1998 when the Sabai Tsho in the upper Hinku river basin burst. The flood peak that reached the Kosi barrage about 300 km downstream was not recognized as an outburst flood until subsequent investigation (Dwivedi, 1999). A crude estimate of the frequency of outburst floods in the eastern Himalaya since 1960 (years/events) shows that glacial lake outburst floods have occurred at an average interval of 3 to 4 years, demonstrating that they are relatively common events.

4. Responses to the Threat of Outburst Floods

All the evaluations of the 1985 Langmoche flood recommended that the government of Nepal and foreign aid agencies conduct inventories of glacial lakes and assess the hazard to existing communities and potential development projects, such as bridges and hydroelectric facilities (e.g., Ives, 1986; Vuichard and Zimmerman, 1987). Several such studies have been completed (e.g., Liu and Sharma, 1988; Mool, 1993; Yamada and Sharma, 1993; Yamada 1998). These studies combined evaluations of lake size, guesses about dam stability, and information about communities and infrastructure at risk downstream to develop lists of lakes considered dangerous. These various assessments have generally regarded four lakes to pose the greatest risks in the near future: Tsho Rolpa in Tamba Kosi river basin, Imja in Dudh Kosi river basin, Thulagi in Marsyandi river basin, and Lower Barun in Arun river basin. A more exhaustive study was completed in September 2000 by the International Centre for Integrated Mountain Development that inventoried 2,315 glacial lakes in Nepal and 2,674 glacial lakes in Bhutan. The study assessed the risks posed by these lakes and identified as "potentially dangerous" 26 lakes in Nepal and 24 in Bhutan.

Following the Langmoche and Chubung floods, the residents of the Khumbu and Rowaling valleys became concerned about the prospects of additional glacial lake outburst floods. By 1992, Sherpa inhabitants of these valleys were expressing concern about the threats to their communities. In May 1992, representatives of almost every household in Beding, the principal village of Rowaling, petitioned the embassies of the major donor nations to Nepal to help avoid an outburst of Tsho Rolpa, the massive lake just 10 km upstream of their village (Damen, 1992). Warnings of the hazard posed by Tsho Rolpa were publicized by Summit Trekking of Kathmandu. Residents of the Rowaling valley continued to solicit help from the governments of Nepal and The Netherlands through the Netherlands-Nepal Friendship Association in 1994. As a result of these requests, an experimental siphon system was installed in July 1995. This small system transmitted lake water at a flow rate of 170 liters per second through a pipe about 15 cm in diameter and 200 m long. This system successfully functioned for more than a year without maintenance, but the outflow pipes separated under snow loads during the winter (Yamada, 1998). In addition to the pilot drainage system, a cooperative project between Nepal and the World Bank installed 19 automatic sirens in 17 different villages in 1997 along the Rowaling and Bhote Kosi rivers to warn of an approaching flood in case the moraine dam of Tsho Rolpa failed.

Because of the perceived urgency to act in reducing the hazard of Tsho Rolpa, alternative methods of drainage were considered, and a decision was made to excavate a channel. The Tsho-Rolpa GLOF Risks Reduction Project is a joint project of Nepal's Department of Hydrology and Meteorology and the government of The Netherlands. This project was able to transport some heavy equipment in pieces via helicopter for reassembly at Tsho Rolpa. A coffer dam was started in December 1998 and completed in July 1999. An artificial spillway was then constructed with completion in June 2000 at a cost of about USD 2.7 million, including about USD 1.1 million in transportation costs. This 700 m long spillway lowers the lake level by three meters, at a rate of 6 $m^3\ s^{-1}$. Lowering the lake level by at least 10 m and possibly up to 20 m is thought to be necessary to reduce the risk of an outburst flood to a modest level (at a cost of another USD 5 million).

In the neighboring Solu Khumbu region to the east of the Rowaling valley, a nongovernmental organization was formed to publicize the hazard of an outburst from the lake of the Imja Glacier in the Sherpa heartland. The original goals of this "Save the Imja Valley Committee" included soliciting ideas, technical assistance, and financial support to address the hazard. An attempt to establish a warning system failed to advance because of inability to raise funding and political obstacles. The initial enthusiasm of this group has faded, and this NGO may be currently dormant. In the mid-1990s, there were wide differences of opinion among the Sherpas of Solu Khumbu about whether to actively do anything about the Imja Glacier lake. Some people who had lost property during the Langmoche flood were quite anxious about another outburst and avoided the valley bottoms during the monsoon, while others were relatively complacent. Still others were worried about the hazard but were concerned that human intervention in a natural process could make matters worse. Some residents fear the consequences of non-Sherpa people agitating the natural forces, and believe that there were lessons from the 1985 flood that destroyed the foreign-built hydroelectric plant but killed only a few residents. The communities of Solu Khumbu have not yet reached consensus about whether to attempt to reduce the hazard of the Imja Glacier lake in the same way that the Rowaling Sherpas largely agreed to solicit outside resources. The decision about whether to attempt any action carries great risks in either direction and is particularly difficult in a risk-adverse society where much of the technical information comes from outsiders.

5. Recommended Next Steps

Now that several inventories and assessments of the glacial lakes of Nepal have been completed and some scientific information has reached the communities at greatest risk, the residents of the threatened valleys, the government of Nepal, and foreign aid agencies need to decide whether and how to proceed in reducing the outburst hazard. The assessments have identified the priority lakes, and there are relatively few. The lake-level reduction work at Tsho Rolpa should proceed as soon as additional funding is available. The Imja Glacier lake threat should be addressed as soon as possible. The matter has been actively discussed for more than a decade, and detailed hazard maps of the downstream areas have been prepared (Braun and Fiener, 1995). Development of foreign funding should be relatively easy because the Mount Everest region is so well known internationally. Education of the residents of the Marsyandi River valley about the Thulagi Glacier lake needs to be accelerated so that decisions there do not require as much time as has already passed in the case of Solu Khumbu residents. In the case of the Lower Barun glacier, the downstream valleys are sparsely inhabited for tens of kilometers and there is not a local constituency concerned about a threat far upstream. Other lakes need continued monitoring to evaluate developing hazards. In general, the risk of an outburst increases with time as the lake grows, the impoundment weakens, and the opportunities for a trigger persist. There is obvious urgency in addressing the problem.

The communities need to make the first steps by seeking and distributing information about the upstream hazard. Although outsiders can catalyze and assist this effort, local leaders must direct the community decision process. A couple of dedicated leaders could make a significant difference in stimulating the decision process and, if chosen, soliciting technical and financial resources. Thorough discussions of the natural risks and those of inaction and alternative actions are necessary. Although the affected communities may choose to continue to live with the hazard, such a choice should be a deliberate decision made with as much information as is available, rather than avoiding or delaying a decision. In the Solu Khumbu area, the Save the Imja Valley Committee could play a critical role in educating people and conducting discussions. Pursuing development of a reliable radio warning system through national and international agencies would be an obvious first step that would probably have widespread support in any community.

Technical agencies should be ready with contingency plans to respond to the wishes of the affected communities. Planning for measures to reduce the risks of an outburst flood can proceed while communities make their decisions, and such planning can contribute to the community debate. The relative risks of action and inaction should be portrayed as clearly as possible to local residents. Several scenarios can be imagined in which the engineering operations could trigger an outburst. The most obvious alternatives for engineering approaches have been discussed by Damen (1992), Grabs and Hanisch (1993), and Kattelmann and Watanabe (1997). However, there is much room for creative thinking. Warning systems should be developed as soon as possible, and can be expected to be endorsed by any community. Agencies should evaluate the system used in the Rowaling valley for application elsewhere. The experience with excavating the canal at Tsho Rolpa also should be thoroughly evaluated for application to the subsequent deeper canal and to other glacial lakes. Technical advisors have a responsibility for communicating the state of knowledge without creating a sense of panic. Apparently during the work at Tsho Rolpa, many downstream residents were led to believe that a catastrophe was imminent. There are lessons to be learned from the social aspects of the Tsho Rolpa experience as well.

Because Nepal lacks the financial capacity to deal with the glacial lake outburst hazard, external funding will be necessary. Some fraction of the foreign aid proposed for various water resources development projects could be directed toward reduction of this water-induced hazard. Billions of US dollars have been spent in recent years on recovery missions following various natural disasters. Here is an unusual opportunity to reduce the possibility of disasters before they happen at relatively low cost rather than reacting to another calamity. Unfortunately, the problem was not addressed during the United Nations' International Decade for Natural Disaster Reduction, though it should have been an obvious candidate project. Perhaps global visibility for glacial lakes can be obtained during the 2002 International Year of the Mountains.

Acknowledgments

W. E. Grabs informed the author of the potential utility of siphons in reducing the hazard of glacier lake outburst floods. T. Watanabe, M. Damen, T. Yamada, P. K. Mool, A. Pokharel, and T. Kadota provided important information about Tsho

Rolpa and Imja Glacier Lake. The Save the Imja Valley Committee can be reached in care of J. P. Lama Sherpa, GFAS Pvt. Ltd., P.O. Box 3776, Kathmandu, Nepal.

References

Benn, D. I., Wiseman, S., and Warren, C. R.: 2000, Rapid growth of a supraglacial lake, Ngozumpa Glacier, Khumbu Himal, Nepal, In: M. Nakawo *et al.* (eds), *Debris-Covered Glaciers*, Publication 264, International Association of Hydrological Sciences, Wallingford.

Braun, M. and Fiener, P.: 1995, *Report on the GLOF Hazard Mapping Project in the Imja Khola/Dudh Kosi Valley*, Nepal, Snow and Glacier Hydrology Project, Nepal, Department of Hydrology and Meteorology and German Technical Cooperation, Kathmandu, Nepal.

Buchroithner, M. F., Jentsch, G., and Wanivenhaus, B.: 1982, Monitoring of recent geological events in the Khumbu Area (Himalaya, Nepal) by digital processing of Landsat MSS data, *Rock Mechanics* **15**, 181–197.

Carling, P. A. and Glaister, M. S.: 1987, Reconstruction of a flood resulting from a moraine-dam failure using geomorphological evidence and dam-break modeling, In: L. Mayer and D. Nash (eds), *Catastrophic Flooding*, Allen and Unwin, Boston.

Damen, M.: 1992, *Study of the Potential Outburst Flooding of Tsho Rolpa Glacier Lake, Rowaling Valley, East Nepal*, International Institute for Aerospace Survey and Earth Sciences, Enschede, The Netherlands.

Deslodges, J. R. and Church, M.: 1992, Geomorphic implications of glacier outburst flooding: Noeick River Valley, British Columbia, *Canadian Journal of Earth Sciences* **29**, 551–564.

Dwivedi, S. K.: 1999, (Outburst of Sabai Tsho – exact title unknown). *Bulletin of the Water and Energy Secretariat*, Kathmandu. (reported in Kathmandu Post, June 17, 1999).

Evans, S. G. and Clague, J. J.: 1994, Recent climate change and catastrophic geomorphic processes in mountain environments, *Geomorphology* **10**, 107–128.

Fushimi, H., Ikegami, K., Higuchi, K., and Shankar, K.: 1985, Nepal case study: catastrophic floods, In: G. J. Young (ed.), *Techniques for Prediction of Runoff from Glacierized Areas*, Publication 149, International Association of Hydrological Sciences, Wallingford, pp. 125–130.

Gansser, A.: 1970, Lunana, the peaks, glaciers, and lakes of northern Bhutan, *The Mountain World* **1968/69**, 117–131.

Grabs, W. E. and Hanisch, J.: 1993, Objectives and prevention methods for glacier lake outburst floods (GLOFS), In: G. J. Young (ed.), *Snow and Glacier Hydrology*, Publication 218, International Association of Hydrological Sciences, Wallingford.

Ives, J. D.: 1986, Glacial lake outburst floods and risk engineering in the Himalaya, Occasional Paper No. 5, International Center for Integrated Mountain Development, Kathmandu.

Kattelmann, R. and Watanabe, T.: 1997, Draining Himalayan glacial lakes before they burst, In: G. H. Leavesley *et al.* (eds), *Destructive Water: Water-Caused Natural Disasters, their Abatement and Control*, Publication 239, International Association of Hydrological Sciences, Wallingford.

Liu, C. and Sharma, C. K. (eds): 1988, *Report on First Expedition to Glaciers and Glacier Lakes in the Pumqu (Arun) and Poiqu (Bhote–Sun Kosi) River Basins, Xizang (Tibet)*, China Science Press, Beijing.

Mool, P. K.: 1993, Glacier lake outburst floods in Nepal,*Water and Energy Commission Secretariat Bulletin* **4**(1), 8–11.

Muller, F.: 1959, Eight months of glacier and soil research in the Everest region, *The Mountain World* **1958/1959**, 191–208.

Oi, H.: 1993, Nepal and disasters, *INCEDE (International Center for Disaster-Mitigation Engineering) Newsletter* **1**(4), 1–4.

Poudyal Chhetri, M. B.: 1999, An overview of disaster management in Nepal: Nepal country report, 1999. Website of Asian Disaster Reduction Centre, http://www.adrc.or.jp/countryreport/NPL/NPLeng99/Nepal99.htm

Pearce, F.: 1999, Flooded out, *New Scientist* 5 June 1999.

Richardson, S. D. and Reynolds, J. M.: 2000, Degradation of ice-cored moraine dams: implications for hazard development. In: M. Nakawo *et al.* (eds), *Debris-Covered Glaciers*, Publication 264, International Association of Hydrological Sciences, Wallingford.

Rothlisberger, H. and Geyh, M. A.: 1985, Glacier variations in Himalaya and Karakorum, *Zeitschrift fur Gletscherkunde and Glazialgeologie* **21**, 237–249.

Vuichard, D. and Zimmermann, M.: 1986, The Langmoche flash-flood, Khumbu Himal, Nepal, *Mountain Research and Development* **6**(1), 90–94.

Vuichard, D. and Zimmermann,M.: 1987, The 1985 catastrophic drainage of a moraine-dammed lake, Khumbu Himal, Nepal: causes and consequences, *Mountain Research and Development* **7**(2), 91–110.

Walder, J. S. and Fountain, A. G.: 1997, Glacier generated floods. In: G. H. Leavesley *et al.* (eds), *Destructive Water: Water-Caused Natural Disasters, their Abatement and Control*, Publication 239, International Association of Hydrological Sciences, Wallingford.

Watanabe, T., Ives, J. D., and Hammond, J. E.: 1994, Rapid growth of a glacial lake in the Khumbu Himal, Nepal Himalaya: prospects for a catastrophic flood, *Mountain Research and Development* **14**(4), 329–340.

Watanabe, T., Kameyama, S., and Sato, T.: 1995, Imja Glacier dead ice melt rates and changes in a supra-glacial lake, 1989–1994, Khumbu Himal, Nepal: danger of lake drainage, *Mountain Research and Development* **15**(4), 293–300.

Watanabe, T. and Rothacher, D.: 1996, The 1994 Lugge Tsho glacial lake outburst flood, Bhutan Himalaya, *Mountain Research and Development* **16**(1), 77–81.

Wilford, J. N.: 1988, Glacier-formed lakes seen as potential catastrophes, *New York Times*, 21 June, 1988.

Yamada, T.: 1993, *Glacier Lakes and Their Outburst Floods in the Nepal Himalaya*, Water and Energy Commission Secretariat, Kathmandu.

Yamada, T.: 1998, *Glacier Lake and its Outburst Flood in the Nepal Himalaya*, Data Center for Glacier Research, Japanese Society of Snow and Ice, Tokyo, Monograph No. 1.

Yamada, T. and C. K. Sharma: 1993, Glacier lakes and outburst floods in the Nepal Himalaya, In: G. J. Young (ed.), *Snow and Glacier Hydrology*, Publication 218, International Association of Hydrological Sciences, Wallingford.

Yamada, T., Shiraiwa, T., Iida, H., Kadota, T., Watanabe, T., Rana, B., Ageta, Y., and Fushimi, H.: 1992, Fluctuations of glaciers from the 1970s to 1989 in the Khumbu, Shorong, and Langtang regions, Nepal Himalayas, *Bulletin of Glacier Research* **10**, 11–19.

Natural Hazards **28:** 155–179, 2003.

Floods and Vulnerability: Need to Rethink Flood Management

AJAYA DIXIT
Nepal Water Conservation Foundation (NWCF), P.O. Box 2221, Kathmandu, Nepal
E-mail: nwcf@wlink.com.np.

(Received: 21 May 2001; accepted: 4 March 2002)

Abstract. Responses to flooding in the Himalaya-Ganga region have conventionally been sought under the hierarchic mode using the strategy of control. Because it has not given due consideration to diverse contexts of the region and thereby the specific approaches that such context necessitate, the particular approach has not brought about security from flooding. Three responses are seen in the terrain of flood disaster. These are the hierarchic manager, individualistic innovator and the egalitarian social activist. The hierarchies define control as the solution to the problems of flood: this is the approach preferred by state agencies. At the individualistic level the sought strategy is flexibility to cope with the situation. The response by social activists is guided by egalitarian critiques of the hierarchic approach. Each pursues his/her own styles and continuously contests the policy terrain. This paper reviews the nature of flood disaster in the Himalaya-Ganga by focussing on plains Nepal. It argues that conventional approach has not been able to provide the security envisaged. The paper suggests that vulnerability of people in risk-prone areas must be addressed by enhancing resilience capacity. For this to happen the approach must be pluralistic that gives space to each management style with varying obligations at varying scales.

1. Introduction

Processes such as landslides and floods are part of the Himalaya-Ganga's diverse geographical make-ups. Nepal is part of the Himalaya-Ganga, the generic name for a highland-lowland interactive system consisting of the highest mountain chain on this planet.* The extreme behaviour of nature ranges from sudden and high intensity monsoon rainfalls, to shifts of continental plates. In the Himalaya-Ganga the variety of topographic and climatic types contains an abundance of ecological niches harbouring widely diverse species of flora and fauna. The pressures of population growth and the needs of modernisation, however, have created a head-on confrontation between the environment and people. Interventions for development need to be effectively synthesised with concerns emanating from the complex nature of the physical environment and its extreme and non-temperate character. The region is shown in Figure 1.

* Verghese (1990) has used the expression to highlight development potential available in the Himalayan rivers. Gyawali and Dixit (1994) define the regional complexity as the Himalaya-Ganga.

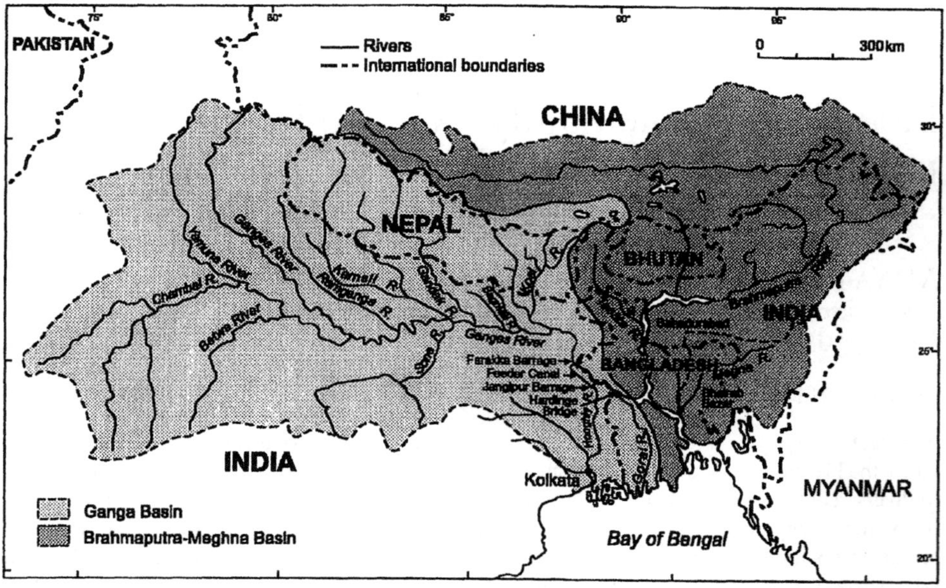

Figure 1. The Himalaya-Ganga Region.

In this region, the monsoon is both a productive resource and a hazard at the same time. With the beginning of the monsoon the situation of no water changes into a situation of too much water. During the monsoon cloudbursts, landslides, mass movements, and flash floods are common in the mountains. When the flow reaches the plains of southern Nepal, northern Uttar Pradesh, Bihar and West Bengal, rivers overflow their banks and inundate the land leading to the large-scale disruption of social and economic lives. As they flow from the mountains to the plains, rivers also cut banks and shift laterally. In the upper regions, rivers cause erosion and transport sediment. The sediment mass thus transported gets deposited in the plains. Fine sediment brought by floods enriches productivity of land. At the same time rivers also bring larger sediments whose deposition on agricultural lands harms productivity. The combined effects constitute a disaster.

With the rise in population, cities have grown. Highways and railways have been built along with flood embankments. Poorly designed and located, these infrastructures have interfered with the existing pattern of surface water flow, and caused drainage congestion. In turn they have exacerbated the flood hazard. Both push and pull factors have caused migration from the hills and rural areas to urban regions, which have increased occupancy on the flood plains. Many times the migrants end up living in a vulnerable zone because they have no opportunity to live in less hazardous area. The rising number of migrants adds to the poor in these regions already living in vulnerable conditions. Combined effects of migration and socio-economic condition have also increased the competition for jobs and resources, which in turn has accentuated social stress among the people. The combined impact disproportionately affects the poorer households – the landless and assets poor farmers.

Considering a suitably long time frame, one realises that flood events are a regular phenomenon in the region. As forecasting techniques for extreme events are not very accurate, coping with disasters has been treated as a question of relief after the event has occurred rather than as a risk that should be anticipated. Flood disaster mitigation involves large commitment of financial resources that could otherwise be spent on mainstream development effort. Disaster preparedness would imply making adequate initial investments in acquiring a better, more scientific understanding of the causes of natural hazards and a parallel effort in improving institutions and infrastructure necessary to confront them. It also means increasing the role and involvement of social institutions that range from local governments to non-governmental actors at the regional and the state levels.

This paper analyses flood disaster in the Himalaya-Ganga focusing on Nepal. Till the early 1950s, a combination of historical isolation and geographical remoteness kept the country secluded from the influences of the outside world. Since then, modern development and technology have arrived in Nepal, and the country has had to face the challenge of natural disasters with modern means, as have her neighbours. While the monsoon links the highland and lowland region with an integrated hydrology, the plate tectonics of the Indo-Asian collision integrates its geology.

2. Physical Context

Nepal's geographical area of 147,181 km^2 is placed at the centre of the Himalaya and is composed of succession of east-west mountain ranges alternating with river valleys. The country is divided into three principal geographic and ecological zones: the southern plains, the middle hills, and the mountains. The plains include the Tarai, which is an extension of the Gangetic plains constituting 17% of the land in Nepal. Stretching along the southern border of the country, the Tarai is almost 300 m above sea level in the Far West, but slopes to just 60 m in the east. The Siwalik Hills range from altitude of 500 m to 1800 m in height while the Mahabharat Range alternates between 1500 m and 2700 m. The intervening Inner Tarai valleys (also called the *Dun* valleys) are generally between 200 m to 1000 m in elevation.

Between the Mahabharat and the High Himalayan range is the middle hill region. Altitudes range from 400 m to 3000 m in this extremely dissected hill region, in which the majority of the population lives. The lowest points in the region correspond to the river valleys, while the higher elevations consist of mountain spurs referred to as *lekhs*. The horizontal distance between them is sometimes only a few kilometres. The mountain region generally lies above 2500 m and consists of the high spurs of the Himalayas as well as trans-Himalayan plateau and valleys. The geology of Nepal is a product of continental plate tectonics and the natural weathering processes, which together created the Himalaya.

The diversity of topographic features that has arisen from plate tectonics and weathering processes has resulted in a heterogeneous distribution of soil types. The Tarai plains consist mainly of deep layers of alluvium built through riverine and geomorphologic activities. Along the foot of the Churia Range the soil is often fairly coarse and intermixed with sand and gravel. The Churia Hills are composed mainly of sandstone, siltstone and shale. In the hills and mountains, soil type and depth depends on the character of the adjoining geology and the prevalent slope. On the valley floors, the soil is deep and has a high degree of organic detritus deposited by floods.

As a result of the ongoing subduction of the South Asian plate under the main Asian plate, the geological faults are still very active; this activity is causing the Himalayas to continue to rise. This "continental collision" result in earthquakes, including major tremors of 1934 and 1988, which struck the eastern Tarai regions of Nepal and India. Such recent examples of high seismic activity contribute to the geological instability of the country's landmass. Compounding this instability is heavy precipitation, the high gradient drainage from the steep mountain slopes, and weathering processes, all of which in turn lead to soil deposition in the lowlands, valleys, and the highland terraces, called *tars*.

The monsoon rain, which lasts from June to September, is marked by large regional and temporal variations. The rainfall is sharp and intense. The date of onset, and the magnitude, duration and intensity of rainfall vary at macro, meso and micro scales. The eastern region is wetter than the west. The amount of rainfall in the east is higher than in the centre and west Himalayas. In the rain-shadow regions of the Tibetan plateau, conditions are dry and desert-like. In the mountains, orographic effects cause large local variations within a valley.

Altitude and topography also condition the climate significantly. There are basically three seasons in the country: the monsoon between June and September, the winter between October and February, and the dry season between March and May. The monsoon accounts for about 80% of the total annual precipitation. Because the summer monsoon advances from the east in Nepal, precipitation is heavier in the east than in the west. The remaining 20% of the precipitation fall in the winter, this rainfall is distributed more in the north and west of the country than in the east. The distribution pattern is due to the fact that this rainfall results from a depression, which originates in the Mediterranean and weakens as it progresses eastward. Meagre as it may be, rain during this season is critical for agriculture in some areas of Nepal. For example, areas a village in the trans-Himalayan region gets approximately 46% of its annual precipitation from the winter source. Much of this falls as snow that melts in spring and contributes to an increase in the dry season flow of snow-fed rivers.

The spatial distribution of rainfall depends upon the nature of orographic profiles. In contrast to the southern slopes of the Churia and the Mahabharat, which receive the highest amount of rainfall, the northern slopes fall in the rain shadow area. On an average, Nepal gets about 1600 mm of rainfall annually. According to the isoheytal maps, certain areas, such as the upper Arun catchment, upper Sun Kosi catchment and the Pokhara valley show annual averages of almost 4000 mm, while the trans-Himalayan areas of Jumla and Mustang register as little as 250 mm. In addition to spatial variations, there are sharp temporal differences from year to year. Cloudbursts can deliver 300–540 mm of rain within a single 24-hour period, which often result in flash floods.

Topography accounts not only for variations in precipitation but also for those in climate. Altitudes between 60–500 m have a sub-tropical climate and the average daily mean temperature ranges from 15 °C in January to 30 °C in May. Maximum temperatures rise to 45°, while the minimum descends to two degrees below zero. The climate between 500–2500 m, varies from warm temperate to cool temperate. The daily means in January and July respectively vary from 10–25 °C. Cold spells often result in temperatures below the freezing point. Areas between 2500–4000 m are characterised with an alpine regime, while at elevations above 4000 m - arctic conditions prevail. In the inhabited villages of the lower mountains altitudes, summer daily mean temperatures rarely exceed 18 °C, while those in winter seldom rise above five degrees.

Although it is the extreme events that cause most damages, they have not been the focus of analyses, which instead focus on averages, which, however, do not present a valid picture of these variations. One of the primary difficulties in establishing firm conclusions about the weather and climatic conditions of various ecological regions of Nepal results from the paucity of hydro-meteorological data. Even with partial stations the average rain gauge station covers only 600 km^2. It is estimated, however, that there should be at least one station every 100 km^2 in order to account for the topographical diversity of the country. In addition, the existing stations have limited historical records as they were established only in the 1960's. Furthermore, these stations are concentrated in the Bagmati and Kosi catchment areas and in the more accessible valleys below 1500 m, rather than in the inaccessible but more representative hills.

In Nepal the monsoon and the Himalayan mountain system give rise to about 6000 rivers of various lengths. The rivers can be broadly categorised into three types; snow-fed, rain-fed originating below the snow line, and rain-fed originating in the southern face of the mountain chain and the plains. The Mahakali, Karnali, Gandak and Kosi Rivers are the main tributaries of the Ganga River and originate in Tibet and the Nepal Himalayas. The Ganga River and its major tributaries also drain the floodwaters into the Bay of Bengal. All rivers flowing from Nepal join the Ganga or its tributaries in Uttar Pradesh, Bihar and West Bengal (Figure 2).

In the four major basins – the Mahakali, the Karnali, the Gandak and the Kosi – rivers are perennial and derive their flow from monsoon run-off and snow-melt.

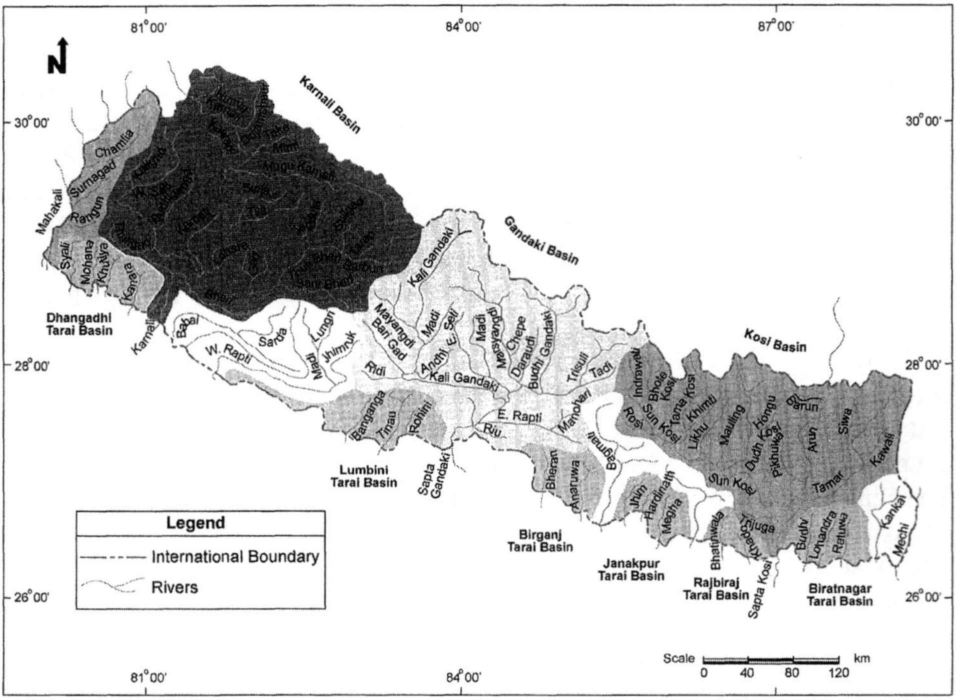

Figure 2. The Ganga and its tributaries.

There are six smaller basins, whose headwaters originate in the middle hills and whose low season flows are not augmented by snow-melt. These are the Babai, the West Rapti, the Bagmati, the Kamala, the Kankai and the Mechi. The East Rapti, the Andhi Khola and the Trijuga have hydrological and morphological characteristics similar to those of the rain-fed rivers. They join the main rivers of the larger basins as tributaries of these larger rivers. The Tarai is criss-crossed by several smaller rivers, which originate in the southern slopes of the Churia Hills and flow southwards into northern Ganga plain. The southern slopes of the Churia Hills receive some of the heaviest and most intense rainfall in the country.

The hydrology of rivers generally follows the pattern of rainfall. From July to September the rivers experience high flow; from October to November a period of decreasing flow ensues. From December to April flow is low. Pre-monsoon rains, snowmelts and thunder squalls increase flow in May. With the commencement of the monsoon in early June, flow starts to increase. The peak monsoon flow of these streams can be a thousand times greater than its lowest flow.

3. Socio-economic Characteristics

Nepal's population of 22.9 million (CBS, 2000) has doubled since 1961 and, at its present growth rate, is likely to double again in the next 35 years. The average growth rate recorded for the 1991–2000 census period was 2.37% (CBS, 2000). Although the figure is lower than that for the previous census decade, there are no overt indications of a significant decline in fertility suggesting that the country is beginning to undergo the demographic transition towards a stable population growth in future. Nepal's population is divided into many ethnic, linguistic and religious groups living in the various ecological niches.

In the absence of significant employment opportunities in the non-agricultural sectors, the growing population, if it does not out-migrate to urban centers in Nepal, India or in recent times to oil rich gulf countries, Japan and Malaysia, is absorbed into the agricultural sector, thereby increasing the stress on the land. Per capita arable land has declined to 0.24 ha in 1990 from 0.6 ha in 1964. The agriculture production index showed an increase of just 44% between 1976 and 1991. This level of production increase is mainly due to the expansion of cultivated land at the expense of forest cover mostly in the tarai.

The Nepali economy is broadly characterized by an low productive agricultural sector, which is mainly subsistence oriented, and a nascent industrial base dominated by agro-processing enterprises. The share of the agricultural sector in total gross domestic product (GDP) and total employment are two of the key indicators of the level of development of a country. On these criteria, Nepal has some of the highest figures even among the poorer countries of the world. The agricultural sector still accounts for over 40% of GDP and over 81% of total employment (CBS, 1991).

After five decades of planned development (1950–1999) the country's social and economic indicators – widespread poverty, low per capita income, high unemployment, subsistence agriculture, low literacy rate and high indebtedness-suggest that Nepal is one of the least well-off countries in Asia.* The sluggish economic base coupled with low education level and lack of access to basic services like health, safe water means poor quality of life. The condition has not changed despite heavy inflow of foreign aid, which in 2000 accounted for almost 65 percent of the development expenditure. The past development efforts has not been able to bring about changes in condition of the people. These conditions and institutional shortcomings mean that the impact of flood disaster falls disproportionately on the asset poor and vulnerable communities.

4. Nature of Flood and Impacts

Floods in Nepal results from the combination of the following natural factors: continuous rainfall and cloudbursts, snowmelt and rainfall combine, glacial lake outburst floods, and *bishyari* (the breaking of dams caused by landslides falling directly into rivers). One major cause of floods is hydro-climatological. At the end

* Nepal leading development thinker has termed this outcome 'failed development' see Dixit (2000) for details. The crisis in providing a macro-economic stability and promoting pro poor governance is discussed in ESP (2001).

of the monsoon, when the land is saturated, the contribution of rainfall to overland flow is almost hundred percent, and cloudbursts can lead to devastating effects. Cloudbursts, which may yield more than 500 mm of rainfall in a single day, are common. In the Kulekhani catchment in 1993, 540 mm rainfall was recorded in 24 hour period. In 1998, the rain gauges installed in the catchment of Rohini in the tarai recorded 3,200 mm of rainfall in the three months of June, July and August, which was almost two times the average rainfall recorded in the nearby stations over the last several years in the same period. On August 3, of 1998 one of the rain gauge stations recorded a rainfall of 459 mm in 24 hours.*

Dhital and Uprety (1994) analysed the frequency of 24 hours rainfall events greater than 100 mm recorded in the country from 1981 and 1990. The frequency of occurrence was plotted and countered on a map. The highest frequency was along the eastern, central and western Tarai of Nepal. Gyawali and Dixit (2001) suggested that high intensity rainfall in the Himalaya-Ganga region is part of the natural cycle necessitating in-depth analysis about their trend and periodicity than available at present. Widespread rainfall has led to major floods in the large river like Kosi. The highest and second highest floods in the river that occurred in 1954 and 1966 were the result of widespread rainfall in its mountainous catchment. In the smaller streams the impact of flooding is more localised, but can become widespread depending upon the magnitude, extent and duration of the accompanying rainfall.

Floods and landslide have to be considered together when their impact in Nepal are analysed. Landslide has been analysed as the geological hazard.** The impact of landslides is severe for local rural hill economies and manifests itself in terms of lost production, destruction of the production base, displacement of households and villages, and the disruption of normal inter-village patterns of movement. With the advent of modernisation and the notion of state responsibility, the direct cost of landslides is compounded further by the additional costs incurred in rehabilitating modern infrastructure such as highway, bridges, canals and suspension bridges.

Rainfall is the major trigger of landslides. Rainfalls also aggravate old landslides because they usually result in heavy and abrasive run-off. Intense rainfall in a single day can loosen steep slopes causing the hillsides to descend into the valleys as mudslide and debris flows. The impact of such slide can be devastating as the 1993 cloudburst showed. The cloudburst triggered a debris flow in the Kulekhani catchment killing 64 persons in Phedigaon alone (DPTC, 1994). Heavy rain by itself is cause enough for high stream flow, which in turn leads to under-cutting of banks and landslides. This phenomenon has a positive feedback cycle in that landslides caused by flash floods temporarily dam rivers, which breach and result

* See Dixit *et al.* (2001) for analysis of the 1998 monsoon events in the Rohini catchment. The three months of June, July and August recorded 3,200 mm rainfall. The normal rainfall in the region recorded earlier during the same three months period was 1,502 mm.

** For a hazard led perspective on landslide in the Himalayan region, see Dhital and Uprety (1994) and Tianchi *et al.* (2001).

in even more intense flooding that trigger even larger landslides, debris flow and mass wasting.

Karmacharya (1989) found that an average of 78 landslides of significant proportions occurred annually in the Central and Western Development Regions of Nepal between 1970 and 1980. The estimate, which also did not include the costs of lost production, rehabilitation or human suffering etc. was Rs 6.5 million per year. This was a highly conservative estimate of the losses, which took into account only the official cost of land and cattle. When this is extended to the other three regions of Nepal and to the remote areas of the country, where such events go unreported, the costs become very significant indeed.

Land slope and fractured rock structures are the main causes of landslides and mass wasting. Intense rainfall, riverbank undercutting and/or groundwater flow, as well as improper slope cutting for road or irrigation canal construction further exacerbate the instability of land. The geological conditions in the fractured hills are area specific and vary in terms of the type of rock and debris, the dip of rocks, the slope of the land, and the extent of groundwater seepage. Consequently, there is considerable variance in the levels of vulnerability to landslides, rockslides, and soil creep, to which the different areas of the Himalaya are susceptible. These instabilities are greatly aggravated by seismic disturbances and take many years to stabilise. Seismic disturbances also lead to erosion and mass movements. According to Goswami (1983) the earthquake of 1950 increased soil creep and slippage over wide areas of Brahmaputra catchment, enhanced erosion and provided a large debris load for the river. From 1953 to 1963 Brahamaputra river showed high sediment load.

Sediment resulting from mass wasting and landslides is often released into river channels, temporarily damming them. When such a dam is breached peak flood of short duration called *bishyari* results. The flood causes heavy damage along riverbanks, and the impact may continue for several kilometres downstream of the breach. The peak is attenuated when the flood wave reaches wider river reaches. The occurrence of *bishyaris* is random and the flood wave is rapid to allow sufficient warning time. One such event occurred in the Tinau River in 1978. A landslide caused by 125 mm of downpour falling within a few hours blocked the river. The subsequent over-topping of the dam wiped away Dauretole of Butwal causing heavy damages to life and property. Another event in the Tamor River in 1980 caused shifting of the Kosi River's channel at Chatara necessitating major revision of the intake of the irrigation system being built in the river (Mahmood, 1987). In September 1998, monsoon rains triggered a landslide of 400,000 cubic metre in the Kali Gandaki river blocking it for seventy two hours (Volk, 2000). The breach of the dam caused major damage locally (Chettri and Bhattarai, 2001).

As they reach the Tarai plains, rivers deposit their coarser sediment load in the "inland deltas" and cause them to move laterally. The shifting tendency is common all rivers. It is most dramatic in the Kosi River, which has shifted 150 km westwards in the last 150 years.* Intermittently, deposited sediment is washed away in large flood events and is ultimately transported to downstream reaches. River discharge and water levels during such event rivers exhibit high variations. Even without *bishyari* water level can rapidly change. In the Kosi River, for example, water level fluctuation can be sudden and high; a rise of 20 to 30 feet in 24 hours in its gorge is common.

In the High Himalayas, GLOF (Glacial Lake Outburst Flood) is a regular phenomenon. There are many glacial lakes blocked by ice dam and terminal moraine of Himalayan glaciers located above 4000 m elevation. According to Mool *et al.* (2001) 2,323 lakes glacial lakes cover 75 km^2 in Nepal. The Koshi river basin contains 1,062 lakes, the Gandaki river basin 338 lakes, the Karnali river basin 907, and the Mahakali river basin in Nepalese territory contains 16 lakes. Glaciers have retreated rapidly in the last half of the 20th century, forming, in many cases, ice-core moraine-flanked lakes of melt water. Occasionally, there is a breach in a moraine dam and a lake empties in a very short time. The result is floods of great magnitude in downstream river reaches. The water carries with it sediment from the moraine dam as well as from the riverbeds and banks gouged out by the flow. The combined action of sudden flooding and debris movement washes away riparian farmland, settlements, infrastructure and many unsuspecting individuals in its wake.** GLOFs also occur at random, but because the location of a glacial lake is fixed, the regions likely to be affected by a potential GLOF are identifiable.

In 1997 the Tsho Rolpa Lake in Dolakha district reached a critical stage. It storage volume was 100 million m^3. To drain the lake, and minimise the hazard the DHM with support from the Dutch government constructed a spillway in the moraine. The US$35 million project was expected to lower the water level by three meters in two years, thus minimising the threat of a breach. The remediation work conducted at the Tsho Rolpa glacier lake was the first of its kind in the Himalayas. While the lowering of the lake by three meters is expected to reduce the risk of GLOF, it is only a temporary solution (Rana *et al.*, 2000). The mitigation measure was a response in the aftermath of the alarm in Nepal in 1997 that the Tsho Rolpa lake would break and would cause massive devastation in its wake.***

The areas, which could be affected by potential GLOFs and *bishyari*, are sparsely populated, and because the number of people killed or hit is small the intensity of the suffering of the affected populace can be masked. Also since the affected population is in the informal sector of the economy, the losses are weakly accounted-for in statistical counts. Floods in the Tarai, on the other hand, are reg-

* WECS (1987).

** Active glacier lakes are located between the Kali Gandaki in central Nepal and the Kanchenjunga in the east. For an elaborate discussion on location, causes, and lessons of past GLOFs, see Gyawali and Dixit (1997), Mool, (2001) and Dwivedi (2000).

*** Gyawali and Dixit (1997) provide comprehensive account of the panic surrounding the Tso Rolpa in 1997. The discussion on the activities taken to lower the water level of Tso Rolpa including the need to lower the level by further 17 m is found in Rana *et al.* (2000).

ular and provide comparatively longer lead-times than those in the hills. In the 1993 monsoon, however, even in the Tarai the flood event of July was sudden. The flood in the Bagmati River in that year reached the maximum recorded level; its suddenness led to the deaths of about 800 persons in the Sarlahi and Rautahat districts of the central Nepal Tarai.*

Anthropogenic activities such as mining and road building have aggravated mass wasting in Nepal. Study by Carson (1986) indicate that erosion from natural processes far outweigh anything that can be attributed to anthropogenic causes. The treatise on this subject by Ives and Messerli, (1989) argues that human intervention in many instances resulted in the rehabilitation and regeneration of degraded land in different sections of the Himalayas, lowering mass wasting and rapid outflows. This is particularly so when the land is converted into terraces. It has also been argued that terraces on hill slopes contribute to soil conservation and slope stability more effectively than reforestations.** Interventions in the flood plain also affect the upstream–downstream hydrological linkages as structures like weirs and barrages leads to aggradation in upstream channels. During high flow they exacerbate the impact of flooding due to the congestion of the drainage channels.

Nepal suffered a major flood disaster in 1993. In the few days between 19 and 23 July, several intense cloudbursts occurred over the mid-hills of the Mahabharat range south of Kathmandu. The headwaters of the tributaries of the Trisuli, Rapti and Bagmati rivers originate from these heights, and the deluge that washed down these mountain slopes wrought havoc along their tributary streams. The Bagmati irrigation barrage was overtopped by a flood surge that resulted in heavy structural damages and loss of life. A 4000-ton boulder that was carried down the normally miniscule Jurikhet rivulet smashed the penstock of the largest hydroelectric plant in the country causing the unified central Nepal power system to lose almost half of its total generating capacity. Major bridges on the Prithvi and Tribhuban national highways that linked the capital city to the out side world were washed away. Landslides blocked many stretches of the roads (Deoja, 2000). The cloudburst brought heavy sediment mass into the Kulekhani reservoir and almost saturated its dead storage volume.

Flood related disasters bring heavy damage in both the formal and the informal sectors although it is the former that receives most of the media and national attention. Every year flood causes death, cultivated fields and irrigation systems built and managed by generations of farmers are washed away together with trails, bridges and other rural infrastructure. Detail about the damage is recorded, since 1983, but the record is incomplete. Table I summarises the impacts due to floods and landslides.

---

* See DPTC (1994).

** For a discussion of role of vegetation, see Gilmore *et al.* (1986) and on impact of terracing on hill slope see Wu and Thrones (1995). A discussion of the debate surrounding The Himalayan Degradation Theory are found in Chapman and Thompson (1995).

Table I. Losses due to flood and landslide

Years	People (nos.)		Livestock	Houses	Affected	Land	Public	Estimated loss
	Death	Injured	Loss (nos.)	destroyed (nos.)	family (nos.)	affected (Ha.)	infrastructure	(million Rs.)
1983 to 1999	5757	987	41839	116017	411144	84799.34	1968	12549
Data base (yrs)	17	13	17	16	12	14	6	17

Source: DWIDP, 2000.

5. Flood Disasters: Dominant and Alternative Definitions

Most works on flood disasters emphasise the significance of natural trigger events that arise irrespective of human activities.* This literature assumes that disasters are departures from 'normal' social pattern and that recovery from disaster means a return to normal conditions. Policy makers, donors, and relief and development agencies treat flood disasters as isolated events that break the continuity of the 'normal' way of life. Most interventions to mitigate disasters are post hoc responses made under the assumption that an emergency support in the form of relief will help overcome the situation of hardship. Such support aims at restoring the situation to what it was before the disaster. Even when a flood disaster affects the same community every year, government, donor, and non-government organisations respond by providing the same relief and rehabilitation measures each time. Relief is the dominant approach championed by the regions' governments, minion of the state, including international agencies and donor agencies.

This approach does not consider the situation of a society during normal times between the occurrences of two hazard events important. Disasters are considered as a coincidence when a hazard interferes with society. Many times the terms "natural hazard" and "natural disaster" are used interchangeably. The focus is on physical events, which explain the hazards, themselves and the effort is to predict their magnitude and occurrence using hydrological and meteorological science. This approach assumes that disasters are inherent characteristics of natural hazards, which are inevitably caused by the impact of hazard on people and their activities. In the hazard-led approach those affected by a disaster are not involved in the decision-making about mitigation, which is top-down and bureaucratic.**

The growing body of recent research, however, shows that disasters are the outcome not only of natural hazards, but also of socio-economic structures and political processes that make individuals and families vulnerable.*** Labelled as the "alternative approach", this perspective focuses on the various ways in which social systems operate to make people vulnerable to disasters. This approach considers that natural hazards are significant and normal characteristics of the regions where

* See Blaikie *et al.* (1994).

** Maskrey (1989) terms this the dominant approach.

*** This approach is also called the "political-economy" approach *Ibid.*

they occur but also recognises that vulnerability is equally important. Vulnerability is the condition of a person or a group, in terms of their capacity to anticipate, cope with, resist, and recover from the impact of a natural hazard.*

In the countries of the developing world the 'normal' daily life of large section of the people is not different from the living conditions of those hit by disasters. Even in normal times people live in vulnerable conditions. Vulnerable individuals and families find it hardest to reconstruct their livelihood following a disaster. Families do not live in conditions that are vulnerable to disasters out of ignorance about the hazards or their erroneous perceptions of risk. Instead, most have little freedom to choose how and where they live. Asset poor families, for example, often have no option except vulnerable locations such as flood plains. They are forced to live there not because land use planning is poor, but because the prevailing agrarian relation in their society, and the processes of social and economic exclusion make them unable to own assets in safer areas.

Disaster is an outcome of both vulnerability and hazard. The later is related to natural events and is conceptualised as the probability that in a given period in a given area an extreme, potentially damaging natural phenomena will affect a given zone.** The magnitude of the phenomenon, the probability of its occurrence and the extent of its impact can vary and, in some cases, be estimated within a certain margin of error. The vulnerability of any physical, structural or socio-economic systems to a natural hazard is its probability of being damaged, destroyed or lost. Vulnerability is not static but a dynamic process that depends upon the social, economic and political contexts that change over time, which will consequently affect the probability of loss. If societal vulnerability is exacerbated the loss due to flooding will increase. On the other hand, strengthening social resilience capacity would reduce vulnerability. The magnitude of a natural hazard and of the vulnerability of all the exposed elements in a determined moment implies the risk of the hazard. Risk is related to disaster as it includes the loss and damage that can be suffered after a natural hazard including dead and injured people, damage to property, and interruption of activities. In other words, risk is a future potential condition likely to happen.

Evidence from a number of case studies points to the fact that the asset poor individuals and families, at the margins of the region's social, economic and political system often suffer from flood disasters. Because of asymmetric socio-economic relationships these families have no access to basic resources, such as land, food and shelter, health and education. The result is low social resilience. These social, political and economic conditions and their interrelationships during 'normal times' determine why certain sections of the societies are more vulnerable to disasters than others. Relationships and structures dictate access to resources of various types, thereby access to assets, social power and capacity to make decisions

* Op cit Blaikie *et al.* (1994).

** See UNCHS (1981).

about their own lives.[*] Disasters are the unresolved problems of societies during 'normal' times, and these problems affect the way people are impacted by disasters. Analysing these processes and organizations would expose these conditions and help identify the root causes of disasters. Successful responses to flood disasters need to consider how and why of those hit by disasters.

Because of the disruptive nature of flood disasters, their study allows for the analysis of social relationships and the possibilities of changing them. During times of disaster, a society is exposed thus allowing non-linear relationships to be examined.[**] In this sense, the study of disasters is a forensic moment[***] and an opportunity to create new arrangements that will allow vulnerable groups and those repeatedly affected by disasters to reduce their vulnerability. Consequently, not just disaster relief but events between disasters, development initiatives and local power structures are important. To that end, strategies must strive to change those patterns of relationships and institutional arrangements that exacerbate vulnerability to disasters. Reforming these relationships lies at the root of strengthening the people's capacity to cope with disasters and should be made the focus of any approach to mitigation. Different groups, families and individuals affected by disaster should be involved in all stages of decision-making and implementation.

These two perspectives on flood disasters bring in technology as the third important factor. The hazard-led approach has focussed on relief and technological measures like dams and embankments to alleviate floods. This focus on technologically oriented solutions, instead of the social, economic and political factors that lead to non-adoption or mal-adoption of technological measures, has aggravated flood problem rather than led to its amelioration.[‡] In India, for example, although engineers expressed doubts about efficacy of technically guided solutions like embankments, they became the chosen path of flood control after the advent of independence. Some levees brought limited relief immediately after their implementation, but their long-term adverse impact has already been felt.

The focus on a broader political-economic outlook is important but must not, however, de-emphasise the need for understanding the specific details of each disaster. Improving our understanding of natural hazards is essential for minimising uncertainty. For example, while dealing with floods, questions like when, where, and why are important. A combined analysis of natural events, technology and their social context would provide a realistic understanding of a flood and how it will impact. Each of the perspective on natural events, technology and society is

---

[*] Social relationship and structures are often consolidated and maintained by institutions, which formalise relationship among individuals; within households and families; among classes and groups; and among individuals, the state and other social organisations, such as private companies and banks. Inequalities between classes, ethnic groups, and gender categories are often a result of, and reinforced by, these social institutions. See *Duryog Nivaran* (1994).

[**] Gyawali and Dixit (1997).

[***] Douglas (1990).

[‡] See Burton *et al.* (1993).

therefore necessary, but by itself none is sufficient. Thus an analysis must separate the contexts into physical, social and other subsystems, through which we begin analysis of natural events as well as of those processes that make individuals and families vulnerable. This is because the impact of a disaster is felt at the individual, family or community levels. Analysis that focuses on hazard, technology and society will capture the interactions among various sub systems and locate the constraints among them in order to be able to design responses that are adapted to a given context.*

6. Response for Mitigation

A major problem is the actual response to flood mitigation, which, conventionally is perceived as the responsibility of the government. There are separate agencies to work on different aspects of floods, ranging from hydrology to relief and rehabilitation, but they are ineffective. There is a conspicuous absence of flood disaster management plans. Even multi-sectoral approaches have had little co-ordination and thereby have been of limited value in mitigating hardships caused by flooding. Despite the fact that flood disasters are widespread and frequent, efforts in Nepal at creating a suitable set of institutions to deal with the associated consequences have been hamstrung by the lack of resources. Its government-centric approach ignored local governments and community based institutions. Even more problematic is the notion that each flood disaster is an isolated event and that post-disaster relief is the only logical response. Disaster is still considered an act of God that strikes once in one location, not something that could happen again and again.

In 1982, the Natural Calamity and Relief Act, which emphasised relief and preparedness as equally important responses to disaster mitigation, was formulated. The Act made provisions for the Central Disaster Relief Committee under the Ministry of Home. Special Disaster Unit was created to function as the secretariat of the committee. The legal provision of the Act however, could not provide break through the traditional understanding of natural disaster (Rana, 1997). Following the 1988 earthquake, the Act was amended in 1989 in order to encompass the preparedness and rehabilitation aspects of disaster management including flood disaster. Along with the Soil and Watershed Conservation Act of 1982, long-term plans were enunciated. A Natural Calamities Assistance Fund was also established.

A subsequent initiative responding to the UN's call for disaster reduction was the preparation of the National Action Plan for disaster mitigation. In 1994 Nepal established the National Committee for the International Decade for Natural Disaster Reduction under the chairmanship of the Home Minister. The problem of co-ordinating implementation emerged as a major limitation and in 1995 the plan was modified. In 1996 the government approved the plan renamed as a National Action Plan on Disaster Management in Nepal. The 1996 plan includes a disaster

* For discussions of the mosaic nature of the diversity and how responses need to adapt to the context see Moench *et al.* (1999).

preparedness action plan, a disaster response plan, a disaster reconstruction and a rehabilitation plan, and a disaster mitigation plan (Rana, 1997). The plan identifies priority items and activities, delineates responsibilities, and stipulates time frame including periods for monitoring and evaluation. The absence of an institutional basis, however, has resulted in poor quality monitoring and consequently in minimal knowledge about disasters. The crux of the problem lies in the central governments unwillingness to recognise local governments and local community groups and its failure to tie in disaster mitigation efforts with the ongoing debate on decentralisation.

Disaster mitigation efforts have received considerable financial and technical aid from donor agencies. The UNDP has initiated a long-term technical assistance programme for formulating a national building code and for introducing improved building materials and construction techniques. The agency has provided support in disaster management with the aim of enhancing the level of preparedness for coping with calamities. The UNDP coordinated the distribution of international relief to the victims of the 1993 floods. Several donor agencies are involved in sectoral programmes and income generation activities in the affected areas because they believe that effective rehabilitation fosters the speedy restoration of the economic base. In recent times, attempts have been made by researchers to document the frequency and occurrence of disasters. With Japanese support the Water Induced Disaster Prevention Training Centre was established which in 2000 was made a full-fledged department. The department publishes an annual water disaster report.

The non-governmental sector, which is active in the developmental arena, has yet to be mobilised comprehensively and effectively for flood disaster mitigation. Agencies such as OXFAM, the Red Cross Society and Lutheran World Service, all of which focus on relief and rehabilitation, are putting emphasis on understanding the linkages between societal vulnerability and natural hazards better. One innovative initiative, completed as a collaborative study of vulnerability to floods along a marginal trans-boundary river flowing from Nepal into India. The study included academics, development agencies, and grass root community groups in Nepal and India. The study has provided insights into the causes of floods, the nature of vulnerability and the type of strategies that would be effective for flood disaster management (Dixit *et al.*, 2001). Empowering the voluntary sector to respond to disaster necessitates both a re-orientation in thinking about disaster and the integration of disaster mitigation into planning processes.

The co-operation of the non governmental sector is critical and should also be solicited to identify factors that exacerbate societal vulnerability, which may vary in type and intensity from one area to another. The fact is that it is not the natural risk of floods, which is always present, but institutional failure, which exacerbates vulnerability. Institutions are conceived to include rules, regulations, practices, laws and organisations of both formal and informal types. Marginal community groups are not able to cope with a flood who face immense suffering. Outbreaks of epidemics, which often follow floods, make the situation worse. The resulting social trauma makes the distress difficult to remedy by only providing relief.

7. Control or Management

The conventional responses to mitigating flooding are of two types. The first response takes the form of post event relief and the second is structural solutions in the form of multipurpose projects and embankments. On plains of north Bihar, for example, more than 3500 km of embankment has been built in the last 40 years. The embankments have resulted in the aggradation of channel beds, and, consequently the spill over of floodwater and waterlogging. The examples of the Kosi and other embankments in the Ganga plain are cases in point (Mishra, 1997). Due to deposition of sediment the beds of jacketed rivers are at a higher level than the adjoining lands. The other solution sought to provide respite from flooding in the lower Himalaya-Ganga region has been the creation of reservoirs with multi purpose objectives in the upstream valleys. Only a few reservoirs have been built, which are mostly for hydroelectricity generation.

Multi-purposes dams have been incorporated in the co-operative agenda of India and Nepal since 1950s. One initiative, the Integrated Treaty on the Mahakali River, which was signed in 1996, sets the stage for building the Pancheswar high dam on the Mahakali River. The other storage project is the dam on the West Seti, which is proposed as a hydroelectric and not a multi purpose project.[*] Both proposals, however, face disagreements over the articulation of rights to regulated waters even though the Mahakali Treaty proposes sharing benefits commensurate with costs.[**] These differences have not been resolved. Even if they are questions about the logic of making investment in large-scale projects remains unanswered. One major question to adopting large-scale water projects with long gestation period is the mobilisation of finances. The preliminary cost of the Pancheswar Project is estimated at $3 billion US. According to the provision stipulated in the Mahakali Treaty, Nepal is expected to provide half of the investment if its claim to half the benefits is to be realised.[***] This investment is about 1.5 times the country's 1998 budget. The macro-economic risk for such a large investment on weal economy of the country has not been carefully examined.[‡]

Pancheswar and West Seti are among the twenty nine reservoirs proposed in the Nepal Himalayas. The reservoirs would yield a gross storage of 100 billion m^3, and a net storage of 61 billion m^3, which when completed, are likely to yield a

---

[*] The Memorandum of Understanding between HMG/Nepal and Snowy Mountain Engineering Corporation of Australia sets stage for this project.

[**] For a elaborate discussion on the politics of cooperative water development see Gyawali and Dixit (1999).

[***] Article 3 of the Mahakali Treaty says, "The cost of the project shall be borne by the Parties in proportion to the benefits accruing to them". See Gyawali and Dixit (2000).

[‡] See Thapa (1997), and Dhungel (1996).

year round regulated flow of about 5000 m^3/s (Water Nepal, 1993).* Each reservoir will moderate flood peaks in its downstream river stretches if the operation of the reservoir is optimised for flood mitigation. In reality such optimisation is not likely to work and provide benefits, as political-economic disincentives would not keep reservoir empty.

The benefits of flood moderation go largely to the rural areas and the government does not get direct revenue. Even benefits from irrigation are a smaller compared to the electricity, which secures the highest revenue. There is also a systemic contradiction: the needs of electricity and irrigation require a full reservoir and those of flood control require an empty reservoir to accommodate high flow. Both objectives cannot be achieved at the same time. Also in the reservoirs built so far flood mitigation benefits has not been achieved. None of the reservoirs built so far in the region except the two reservoirs in the DVC project in West Bengal possess flood-cushioning capacity. In both reservoirs, however, the original capacity was whittled down to accommodate the concerns for electricity generation.**

While these reservoirs are looked upon as the only means to convert the Himalayan water into revenue generating hydroelectricity, during the monsoon season print and electronic media in Uttar Pradesh and Bihar report that the reservoirs need to be built to hold the high flow securing the plains from flood disasters. Following the 1998 floods in Eastern Uttar Pradesh, the media mentioned that specified cubic meters of water was released from dams located in Nepal and the Himalayan Kingdom.*** In September of the same year the Indian Prime Minister visited Gorakhpur, and remarked that "the rivers coming from Nepal are the main cause of the flood devastation in India. The talks with Nepal to build dams are ongoing".‡ Similar reports also appeared in the media in Bihar. The story was repeated in 1999 and 2000.‡‡

These posturing however do not heed to three fundamental factors. The first is that floods and inundation are part of the region's eco-hydrology and that natural events are politically secular. The second is that embankments and intervention in

* There are two storage sites in the Mahakali Basin, seven in the Karnali, seven in the Gandak, five in the Kosi, and nine in other river basins (Gyawali, 2001).

** For discussion, see Dixit (2000).

*** Janeswar Mishra, the former Water Resources Minister of Government of India, visited Gorakhpur and remarked "Indian government should discuss with its Nepali counterparts about the release of waters from that country into India". Agency Reports (1998).

‡ See The Hindustan Times (1998). The emphasis on using dams in Nepal to alleviate flooding in the Northern Ganga plain is a theme common to all political parties. The Kosi High Dam at Barahachetra in Nepal is perceived as the social, economic and hydrological panacea for the state. See Dubey Binod who quotes former minister for water resources Government of Bihar Jagdananda Singh in The Hindustan Times February 15, 1999 who said "Nothing but the high dam (Kosi) can solve our problem". Environmental groups in Bihar counter this point of view. They argue that the high dam will bring unmitigated disaster. See *Barh Mukti Abhiyan* (1997).

‡‡ Gyawali and Dixit (2001) pleaded for bringing in good science that recognises water in its multiple forms – precipitation, surface and groundwater flows – must be managed conjunctively. Drainage capacity of water must be enlarged rather than constricted.

the flood plains have exacerbated drainage congestion, seepage and waterlogging.* The third is high sedimentation. In the monsoon the rivers not only flow water but also very high sediment load, which pose major risk to reservoir due to deposition. Sedimentation has remained a blind spot in the water-led development discourse in the region.** Himalayan rivers in the monsoon obtain solid mass derived from landslide, bank cutting and debris flow phenomenon that have geological rather than anthropomorphic roots. Suspended sediment is only a fraction of the entire mass of solid matter that flows in rivers. The total load consists of percentage of bed loads, but this contribution is not accounted for in sediment budgeting.

The Kulekhani reservoir lost its dead storage capacity just 13 years after impoundment began even though it was expected to have a 100-year life span. The debris flow into the reservoir triggered by the 1993 cloudburst filled a large percentage of its dead storage. In less than 24 hours, the Kulekhani catchment received 540 mm of rainfall and the water level of the reservoir rose by 20 m in nine hours. In the same period 4.8 million cubic meters of sediment was deposited in the reservoir.*** The project's designer had assumed that annual sedimentation rate would be 88,200 m^3.‡ When compared with the actual rate this was a gross underestimate. Though the river had no data on sediment load, the designers categorically suggested that sedimentation would not be a problem. Gyawali and Dixit (2001) has analysed institutional imperatives of the designers suggesting that some institutional filter that selects facts to suit its interests is at work.

Another challenge is related with rehabilitation of flood affected victims. Voluntary groups working in flood mitigation in Nepal suggest that the number of people getting displaced by riverbank erosion is increasing. The first rehabilitation efforts started after the incessant rainfall and landslides of 1954. Subsequent efforts were geared toward voluntary resettlement in the Tarai. Presently, government's support to landslide and flood affected victims is limited to provision of tarpaulin and immediate relief. Some effort was made at rehabilitating the victims of 1993 flood at Sarlahi and Hetauda. There is no organisation in the country with the mandate to rehabilitate displaced people and the country's capacity in this respect is limited.‡‡

* The 1991 State of the Environmental report by CSE begins a systematic documentation of the problem due to embankments. Mishra (1997) marshals evidence that embankments have exacerbated flood hazard in the lower plain of north Bihar. Gyawali (1998) has provided theoretical analysis of hierarchical and egalitarian ways of organising and how different solidarities perceive responses to the flooding in the north Bihar.

** Most of the writings that highlight the potential of water led development make no mention of the risks posed by sedimentation in the Himalayan Rivers. The mainstream discourse of South Asia does not acknowledged sedimentation as a risk. A recent treatise that outlines the framework for sustainable development of the Ganga, Brahmaputra, Meghana basins has nothing to say about the high rate of natural sedimentation of the Himalayan rivers. See Ahmad (2001).

*** Nippon Koei (1996).

‡ The World Bank (1975).

‡‡ For a discussion of the challenges of water project-related dislocation in Nepal, see Dixit (1994).

This limitation also impinges on the proposal to build reservoirs and raise questions about equity.

Reservoirs will permanently submerge villages, agricultural land and displace hundreds of thousand of people of hills of Nepal. Unfortunately, displacement is not an issue that had been recognised in the bilateral co-operative initiative on water development* and is recently being mentioned (Ahmad *et al.*, 2001), thanks to the alarm raising egalitarian social auditors. In the discourse on water-led development in Nepal, the social cost of intervention first began to be written about in 1989. Land is scarce in Nepal and its Tarai has been an obvious site for resettlement, which has often been accomplished at the cost of forest cover. In any case the Tarai, which occupies 17 percent of the country's area already, accommodates 53% of the population. This pattern of resettlement cannot continue indefinitely into the future and alternative innovative solutions will have to be sought to deal with the problem of rehabilitating victims of floods.

8. Rethinking Flood Mitigation: Risk Management

Floods are a reality of the South Asian physical environment and hydrological asymmetry is a given. The dominant response to flood mitigation has been hierarchic, guided by the incentive to control, and it has meant that efforts to mitigate flood disasters are sought in structural solutions like embankments and/or multi-purpose high dams. The structural approach, however, has not been effective. Embankments have greatly exacerbated flood hazards and none of the completed multi-purpose projects have been provided with a cushion for accommodating flood flow despite popular posturing about their flood mitigation benefits, even if the main purpose might be power generation. That the appropriate response to flooding lies in providing unhindered drainage is an intellectual blind spot because of the political-economic incentives of structural flood control measures.

Another constraint is associated with the dissemination of information, facts and knowledge. Because of its role in shaping public perception, the role of the media is central. In the absence of readily available information about the causes of floods, the media tends to highlight negative attributes and build up hype even if the scientific foundation is illogical. Local media in Uttar Pradesh and Bihar in 1998 reported that Nepal opened gates of dams in its territory. This reporting was incongruous because other than the Kulekhani dam, there is no other dam in the upstream region. Myths and incorrect science also sway statements by political leaders. Following the flood of 1987, the government of Bihar had argued, "until the time big dams come up in Nepal Himalayas, it will be forced to stick to embankment construction as a measure of flood control". Activist groups in Bihar questioned such erroneous reporting in the media. The press statement by one

* Pokharel (1988) provided the first account of involuntary displacement in Nepal. Subsequently, Gurung (1989) reviewed the lessons related with the country's three water related involuntary displacement.

group highlighted that there were no dams in the Himalayan Kingdom except in the Kulekhani river. The responses suggest that critique of the dominant paradigm by egalitarian activists transcend the political boundary of the nation state. The group questioned the notion of transferring blame highlighted that root cause of failure was institutional failure.

Transferring blame by propagating myth is not uncommon in the region's environmental politics.* The Theory of Himalayan Degradation of the 1970s, for example, attributed lowland flooding to the hill farmers' cutting trees. Despite findings, which indicate otherwise, this conceptual lacuna is pervasive in academic group, policy making and even the media on reporting in Uttar Pradesh and Bihar showed. Thus, playing the external factor has always been the preferred approach in dealing with flood mitigation in the plains and has allowed mellowing dissent arising out of failures on institutional front.** The notions couched in the statements of politicians and in media reports are manifestations of the political-economy that need to be changed if mitigating the impacts of flooding is to be more responsive to the issues of livelihoods and sustenance.

For the contiguous region along the rivers flowing from Nepal, to Eastern and central Uttar Pradesh, Bihar and West Bengal, the 1998 floods has provided far-reaching lessons. The region will continue to experience high rainfall events albeit with varying magnitude. The extreme floods result from frequent but poorly understood cloudburst during the monsoon. The dams proposed in rivers of Nepal to provide flood cushioning – the Panchesawar on the Mahakali, the Chisapani on the Karnali, the Kosi High Dam at Barahchetra, Bagmati, Kamala (Chisapani) and Kankai etc. – are located in gorges. Even if flood space is optimised in these projects, they will not offset flood disaster to the regions that drain the Churia. The catchments of the Churia rivers would continue to remain unregulated because there is no site in the rivers where a storage dam could be built to store excess monsoon water.

Lack of scientific understanding of cloudburst causes major difficulty. Dixit *et al.* (2001) found that cloudbursts in the tarai section of the northern Ganga plain were the reason for the unprecedented flood disasters in the Rohini and its tributary rivers. Though such events are common, inadequate network density of monitoring gauge in the region means that specific event will remain unrecorded. A minimum requirement for improving flood forecasting is a fair number of rainfall and flow monitoring stations.

The existing hydrological stations provide data of varying levels of reliability for only a few rivers and are inadequate. The establishment of a sound mechanism for data collection, analysis and interpretation should be intertwined with the establishment of a system for floods forecasting. Forecasting of floods need to ensure that information is made available to local people in a language that they can understand. The value of forecasting depends greatly on the accuracy

* Chapman (1995).

** *Ibid.* CSE (1991).

(this affects its credibility), the effectiveness of the message delivery system (Bari, 1995). The delivery and receipt of the message about the flood events cannot be taken for granted. In the case of the 1998 Rohini floods both delivery and receiving mechanisms were not available to the affected families.

The problem is not of drought or floods but of risks, which the farmers and the asset poor families face. While relief and rehabilitation are crucial from the point of view of supporting the households and families hit by floods, equally important are provisions such as risk and life insurance, and, in the case of agriculture, crop failure and livestock insurance.* To overcome the challenges we need to focus on governance because making such provision need to be part of regular governmental functioning. Responses to ameliorate the impact of flooding must be made part of the daily operation, rather than *post-facto* relief. The institutional capacity to respond to floods whenever and wherever they occur must be strengthened.

The ecological and physical interconnectedness of the region calls for improved collaboration between countries even though the flood management at root will be a national endeavour. National efforts need to maintain synergy with those at bilateral and regional levels. Studies of the monsoon, determining its dynamics as it shapes the extent of rainfall events would be good point to make collaborative beginning. Translating key findings of the studies and making them available to local communities in a language that they understand would help maintain dialogue necessary to enhance local capacity. Lead-time available for preparedness and evacuation after the warning is important. Evacuated families in many communities need to be provided with support such as food, medicine, shelter and even drinking water. Therefore, we need to conceive of flood disaster in its entirety and keep people at the centre. Blaikie *et al.* (1994) capture this imperative succinctly:

> To protect people and their homes there needs to be investments in raised housing sites including elevated schools and health posts. To protect their health (and thereby indirect livelihoods) there needs to be protected water supplies for drinking, and primary health care. To protect their livelihoods there needs to be access to and, replacement for land lost through erosion, compensation for the losses of animals and other productions assets. Where necessary, there must of course also be flood prevention measures, but based on an assessment of what actually is needed to reduce vulnerability rather than some grand design which supposedly prevent all flooding without considering who benefits, and who loses, who pays for it, and whether it is needed any way.

For this to happen the approach to flood mitigation must allow forces of entrepreneurial innovation, egalitarian caution and governmental regulation their due institutional participation.** The governments need to focus on the formulation of

* *Ibid.* Chapman (1995) who terms such approaches flexible, which the farmers have done for millennia but with an added institutional extra layer, meaning that the response is local but plugged into the national and regional framework.

** This concept is based on Gyawali and Dixit (2001) *ibid.*

regulations, reviews, and inspections: on the creation of buffer stock to meet needs in case of flooding in isolated pockets, and on monitoring. The focus must be on improving livelihood conditions of the poor between periods of disasters. Strengthening local bodies through political and legal empowerment will increase extent of impartial regulatory functions such as formulating rules, policies and mechanisms for adjudication. New methods and tools have to be devised and internalised in the normal operation of societal and governmental activities. It here that the voluntary sector, as the intermediary between the state and the community, should have a more active role in the mitigation of the impact of floods.

The approach also needs to create built-in system of checks and balances, allow space for lateral thinking, and provide the society with the pluralism in innovation, caution and regulation it needs in order to bring about stable transition in the management of the risks due to flooding. We must continuously focus on reinvigorating our approaches during consecutive disaster events rather than after the events. Otherwise the future is further depletion of social resilience, a rising gap in the power relation and overall societal degeneration.

References

Ahmad, Q. K., Biswas, A. K., Rangachari, R., and Sainju, M. M., (eds): 2001, *Ganga-Brahmaputra-Meghna Region: A Framework for Sustainable Development*, The University Press Ltd., Dhaka, Bangladesh.

Barh Mukti Abhiyan: 1997, Proceeding of the Second Delegates Conference 5–6 April in Nirmali Supaul published by Rameswar Singh for *Barh Mukti Abhiyan.*

Bari, M. F.: 1995, Opportunities and Approaches for Regional Cooperation for the Development of an Alternative Flood Warning System, Regional Cooperation on Flood Warning, Bangladesh Disaster Preparedness Forum, *Duryog Nivaran* and Department Of Water Resources Engineering, Bangladesh University of Engineering and Technology (BUET).

Blaikie, P., Cannon, T., Davis, I., And Wisner, B.: 1994, *At Risk: Natural Hazards, People's Vulnerability, and Disasters*, Routledge, London.

Bruninjzeel, L. A. and Bremmer, C. N.: 1989, *Highland-Lowland Interactions in the Ganges Brahmaputra River Basin*, Occasional Paper 11, ICIMOD, Kathmandu, Nepal.

Burton, I., Kates, R. W., and White, G. F.: 1993, *The Environment as Hazard*, The Guilford Publication, Inc.

Carson, B.: 1985, *Erosion and Sedimentation Processes in the Nepalese Himalaya*, ICIMOD Occasional Paper No. 1., ICIMOD, Kathmandu, Nepal.

CBS: 1991, *Central Bureau of Statistics*, HMG/N, Kathmandu.

CBS: 2000, *Central Bureau of Statistics*, HMG/N, Kathmandu.

Chapman G. P.: 1995, Environmental myth as international politics: The problem of the Bengal Delta, in G. P. Chapman and M. Thompson (eds), *Water and Quest for Sustainable Development in the Ganges Valley for Discussion on the Politics of Environmental Myths in the Ganges River System*, Mansell.

Chhetri, M. B. P. and Bhattarai, D.: 2001, *Mitigation and Management of Floods in Nepal*, Ministry of Home Affairs, His Majesty's Government of Nepal, Kathmandu.

Chitkara, S. C. and Mukhopadhyaya, T. K.: 1999, Flood forecasting: A non structural measure or flood management, *Himganga* **1**(3), April–May–June Patna, India.

CSE: 1991, *Floods, Flood Plains and Environment Myths*, State of India's Environment A Citizen's Report, Centre for Science and Environment, New Delhi.

Deoja, B. B.: 2000, Mountains road development in Nepal: Engineering geological concerns, *Journal of Nepal Geological Society* **22**, 167–178.

Dhungel H.: 1996, Macro-economic Adjustments to Large Energy Investments in a Small Controlled Open Economy – A Policy Analysis of Hydropower Development in Paraguay; unpublished Ph.D. dissertation, University of Pennsylvania.

Disaster Review: 1996, HMG/N, Ministry of Water Resources, Water Induced Disaster Prevention Technical Centre and Japan International Co-operation Agency.

Dixit, A.: 1994, Water projects in Nepal: Lessons from displacement and rehabilitation, *Water Nepal* **4**(1), 74–85.

Dixit, A.: 1999, Reconceptualising Mitigation, *Seminar 478*, Malvika Singh, New Delhi.

Dixit, A.: 2000, Water as an agent of social and Economic Change in Nepal, In: Peter Mollinga (ed.), *Water for Food and Rural Development Approaches and Initiatives in South Asia*, Sage India, New Delhi.

Dixit, A., Dahal, N., and Adhikari, S.: 2001, *Reconceptualizing Flood Mitigation in South Asian Marginalized Rivers: Study of Rohini River*, Nepal Water Conservation Foundation, Kathmandu. (Draft).

Douglas, M.: 1990, Risk as a forensic resources, Daedalus, *Journal of the American Academy of Arts and Sciences* **119**, 1–16.

DPTC, 1994: *Annual Disaster Review*, Water Induced Disaster Prevention Technical Centre, Kathmandu.

DWIDP, 2000: Annual Disaster Review, Department of Water Induced Diseaster Prevention, Ministry of WR, HMG, Nepal.

Duryog Nivaran, 1994: ITDG, Sri Lanka.

Dwibedi, S, K., Acharya, M. D., and Simrad, R.: 2000, The Tam Pokhari Glacier Lake outburst flood of 3 September 1998, *Journal of Nepal Geological Society* **22**.

ESP: 2001, Pro Poor Governance Assessment Nepal Enabling State Programme, Enabling the State Program, Kathmandu.

Gilmore, D. A., Bonnell, M., and Cassells, D. S.: 1986, The effects of forestation on soil hydraulic properties in the Middle Hills of Nepal: a preliminary assessment, *Mountain Research and Development* **7**(3), 239–249.

Goswami D. C.: 1983, Brahamaputra River Assam (India) suspended sediment transport, valley aggradation and basin denudation, Ph.D., University of Michigan, USA.

Gurung, H.: 1989, Policy review and experience of project related resettlement in Nepal, *Water Nepal* **1**(4), 3–5.

Gyawali, D. and Dixit, A.: 1994, Himalaya-Ganga Contending within complexity, *Water Nepal* **4**(1), 1–6.

Gyawali, D. and Dixit, A.: 1997, How distant is Nepali Science from Nepali Society? Lessons from the Tsho Rolpa GLOF Panic, *Water Nepal* **5**(2), 5–43.

Gyawali, D. and Dixit, A.: 1999, Mahakali impasse and Indo-Nepal water conflict, *Economic and Political Weekly* **XXXIV**, 533–564.

Gyawali, D. and Dixit, A.: 2000, Mahakali impasse: A futile paradigm's bequested travails, in Dhruba Kumar (ed.), *Domestic Conflict and Crisis of Governability in Nepal*, Centre for Nepal and Asian Studies T.U., Kathmandu.

Gyawali, D. and Dixit, A.: 2001, Water and science: hydrological uncertainties, development aspirations, and uningrained scientific culture, *Futures* **33**, 689-708.

Gyawali, D. and Dixit, A.: 2001, Appeal for Good Science, *The Kathmandu Post*, August 13th, Kathamndu.

Gyawali, D.: 2001, *Water In Nepal*, Himal Books, Kathmandu.

Gyawali, D.: 1998, Patna, Delhi and environmental activism: Institutional forces behind water conflict in Bihar, *Water Nepal* **6**(1), 67–115.

Ives, J. D. and Messerli, B.: 1989, *The Himalayan Dilemma*, Routledge, London.

Karmacharya, M.: 1989, Landslides in Nepal – In the Period 1970–1980, Tribhuvan University, Kathmandu.

Maskrey, A.: 1989, *Disaster Mitigation – A Community Based Approach*, Oxfam, London.

Mahmood, K.: 1987, Reservoir Sedimentation: Impact, Extent, and Mitigation, World Bank Technical Paper No. 71, World Bank, Washington, D.C.

Mishra, D. K.: 1997, The Bihar flood story, *Economic and Political Weekly* **XXXII**(35), 2206–2217.

Moench, M., Caspari, E., and Dixit, A.: 1999, *Rethinking the Mosaic: Investigation into Local Water Management*, Nepal Water Conservation Foundation, Kathmandu.

Mool, P. K., Bajracharya, S. R., and Joshi, S. P.: 2001, *Inventory of Glaciers, Glacial Lakes and Glacial Lake Outburst Floods: Monitoring and Early Warning Systems in the Hindu Kush-Himalayan Region Nepal*, International Centre for Integrated Mountain Development (ICIMOD), Kathmandu.

Nippon Koei: 1994, *Master Plan on Sediment Control for Kulekhani Watershed*, Nippon Koei, Toyko.

Pokharel, J. C.: 1988, Population displacement by the Kulekhani hydro electric project, some lessons for compensation planning, *Prashashan* **51**, 7–13.

Rana, S. J. B.: 1997, National Action Plan on Disaster Management in Nepal: A Commitment to Natural Disaster Reduction in Disaster Review, DPTC and JICA, Kathmandu.

Rana, B. B., Shrestha, A. P., Reynolds, J. M., Aryal, R., Pokharel. A. P., and Burathoki, K.P.: 2000, Hazard assessment of the Tsho Rolpa Glacier Lake and ongoing remediation measures, *Journal of Nepal Geological Society* **22**, 563-570.

Thapa, P. J.: 1997, Water-led development in Nepal: Myths, limitations and rational concerns, *Water Nepal* **5**(1), 35–57.

World Bank: 1975, *Nepal Appraisal of the Kulekhani Hydroelectric Project*, Report No 833a-NEP November 25, Power Division South Asia Projects Department.

Tianchi, L., Chalise, S. R., and Upreti, B. N.: 2001, *Landslide Hazard Mitigation in the Hindu Kush-Himalayas*, International Centre for Integrated Mountain Development (ICIMOD), Kathmandu.

UNCHS: 1981, *Settlement Planning for Disasters*, UNCHS, Nairobi.

Uprety, B. N. and Dhital, M. R.: 1994, *Landslide Studies and Management in Nepal Himalayas*, paper submitted to ICIMOD, Kathmandu.

Verghese, B. G.: 1990, Waters of Hope – Himalay-Ganga Development and Cooperation for a Billion People, Centre for Policy Research, New Delhi.

Volk, H. R.: 2000, The 1998 Tatopni Landslide in the Kali Gandaki Valley of Western Nepal: case and relation to mass rock creep, *Journal of Nepal Geological Society* **22**, 405–412.

Water Nepal: 1993, *Flood Mitigation by Multipurpose Reservoir* **3**(2–3), 81–82.

WECS: 1987, *Erosion and Sedimentation in the Nepal Himalaya, An Assessment of River Processes*, HMG, Kathmandu.

Wu, K. and Thrones, J. B.: 1995, Terrace irrigation of mountainous hill slopes in the middle hills of Nepal; stability or instability, In: G. P. Chapman and M. Thompson (eds), *Water and Quest for Sustainable Development in the Ganges Valley for Discussion on the Politics of Environmental Myths in the Ganges River System*, Mansell, U.K.

Natural Hazards **28:** 181–198, 2003.

Regional Cooperation in Flood Management in the Ganges-Brahmaputra-Meghna Region: Bangladesh Perspective

Q. K. AHMAD and AHSAN UDDIN AHMED
Bangladesh Unnayan Parishad (BUP), P.O.Box 5007, New Market, Dhaka 1205, Bangladesh (E-mail: bup@citechco.net)

(Received: 15 May 2001; accepted: 4 March 2002)

Abstract. Bangladesh is known to be highly vulnerable to floods. Frequent floods have put enormous constraints on its development potential. Unfortunately, the frequency of high intensity floods is on the rise. So far the country has struggled to put a sizeable infrastructure in place to prevent flooding in may parts of the country with limited success. In recent times, it was found that losses of lives and valuable assets could be significantly minimized by implementing non-structural measures including the improvement of flood forecasting and warning system. The existing flood forecasting and warning capacity of Bangladesh could be more effective if real-time data could be acquired from upstream areas within the Ganges-Brahmaputra-Meghna (GBM) catchment, where runoff is generated. In order to do so, Bangladesh needs to foster an effective regional cooperation with the other GBM regional countries of India, Nepal, and Bhutan. This article examines how GBM regional cooperation could be useful towards managing floods in Bangladesh in particular and the region in general.

Key words: Flood management, climate change, sea level rise, structural and non-structural measures, Ganges-Brahmaputra-Meghna (GBM) region, and regional cooperation.

1. Introduction

Flood is a recurring phenomenon in Bangladesh. It is now understood that catastrophic floods have major adverse consequences for development of the country. Unfortunately, this country is likely to face more devastating and frequent floods in the future, especially under the threats posed by projected climate change and sea level rise (global warming of 1.4° to 5.8 °C, on average, and sea level rise of 9 to 88 cm, on average, by 2001 are in prospect according to IPCC Third Assessment Report), which would increase flood risks, among other consequences, in temperate and topical Asia. And, given its location and climatic conditions, Bangladesh is at the forefront of the projected climate change and sea level rise and their consequences, particularly increased flood risks, both in terms of frequency and duration. This flood related outlook also generally applies to other Ganges-Brahmaputra-Meghna (GBM) regional countries, viz. India, Nepal, and Bhutan (IPCC, 2001).

Despite having meager investment capabilities towards mitigating floods, the country attempted to implement a number of structural flood mitigatory measures with very little success. In addition, it has given due emphasis on non-structural measures including the issuance of flood forecasting and warning. During the recent devastating floods of 1998 and 2000 it was observed that flood forecasting with a good lead time has the potential to save lives and the property of millions of people. To install a good flood forecasting and warning system one requires timely data from the entire catchment area. Since over 93 per cent of the catchment areas of the major rivers flowing through Bangladesh lie outside its borders, there is a strong need for cooperation for timely data sharing between co-riparians. The following sections attempt to provide rationale for such a regional cooperation.

2. Bangladesh: A General Overview and the Hydro-meteorological Setting

Bangladesh is a South Asian developing country located between 20°34′ to 26°38′ North latitude and 88°01′ to 92°4′ East longitude with an area of 147,570 km^2. It is bordered on the west, north and east by India, on the south-east by Myanmar and on the south by the Bay of Bengal. A network of rivers consisting of the Ganges, the Brahmaputra, and the Meghna and their tributaries and distributaries criss-crosses the country. The whole country consists of low and flat land formed mainly by the Ganges and the Brahmaputra River systems except for the hilly regions in the north-eastern and south-eastern parts.

About 80 per cent of Bangladesh is composed of the floodplains of these rivers with very low mean elevation above the sea level; the rest of the country is made up of hills and elevated lands. Only in the extreme north-west land elevations exceed 30 meters above the mean sea level (MSL). There exist two uplifted land blocks, known as the Madhupur and the Barind tracts, with elevations between 1 and 5 meters above the adjoining floodplains. In some places, however, they reach up to 25 meters higher than the adjoining floodplains. Hills along the northern and eastern borders of the country have elevations ranging from 10 to 1000 meters above MSL. In general, these hills have very steep slopes; certain areas have moderate or gentle slopes.

As in other parts of the Indian subcontinent the climate of Bangladesh is highly influenced by the monsoon. It originates over the Indian Ocean and carries warm, moist, and unstable air. The climate is, therefore, humid and warm, tropical in nature; and it is fairly uniform throughout the country. There are three main seasons:

(i) a hot summer, characterized by relatively high average temperature (it rises up to 40 °C for 5 to 10 days in the western part of the country), a very high rate of evapotranspiration, and erratic but occasional heavy rainfall from March to June,

(ii) a hot and humid monsoon with heavy rainfall from June to October, with about 80 per cent of the mean annual rainfall occurring during this time, and

(iii) a relatively cooler and drier winter from November to March, with the maximum temperature ranging between 20–40 °C and the minimum occasionally falling to around 5 °C in the north; frost is extremely rare.

The moisture-laden south-west monsoon heralds the onset of the rainy season. The mean annual rainfall varies widely by geographical location within the country, ranging from 1,200 mm in the extreme west to 5,800 mm in the east and north-east; the average rainfall is about 2,200 mm (MPO 1991). The surface water system of the country is dominated mainly by the three major river systems, the Ganges-Padma, the Brahmaputra-Jamuna and the Meghna (GBM). Figure 1 shows the major rivers of Bangladesh.

Of the total GBM regional surface area, about 7 per cent belongs to Bangladesh; and the GBM river systems discharge about 142 thousand cubic meters of water per second into the Bay of Bengal at peak periods (Rahman and Chowdhury, 1998). On average, out of an estimated total availability of 1350 billion cubic meters (BCM) of water, 1160 BCM (net of the loss through evaporation, evapotranspiration and deep percolation) is available for use in Bangladesh or flows to the sea. This vast outflow is second only to that of the Amazon system. (Ahmad, *et al.* 2001; Rasheed, 2001).

There is a large regional influence on the surface water system of Bangladesh. Although the country's share of the total GBM catchment area is only about 7 per cent, being the lowest riparian it drains most of the water generated in the river systems. Drainage areas of the GBM basins are shown in Table I.

Moreover, about 80 per cent of annual total quantity of 1350 BCM of water is generated during the monsoon months, June–October. (Ahmad *et al.*, 2001). Cherapunjee, the wettest place on earth is located within the GBM catchment area, which contributes significantly to the freshwater flow in the Brahmaputra basin. Also, an average annual sediment load of 0.5 billion to 1.8 billion tons passes through Bangladesh rivers, causing a gradual rise of the river-beds and reducing their ability to discharge the flow quickly. Due to these interacting hydrometeorological and physiological factors, the river systems of the country fail to drain all the excess water during peak monsoon periods; and, on an average, 23–30 per cent of the land faces annual flooding of different extents.

Apart from surface water, groundwater is extensively used for both domestic and agricultural purposes. There is a fairly extensive aquifer at a very shallow depth of 20–40 feet below ground level. A deeper aquifer at about 200–400 feet depth has also been identified in many parts of the country and is used for irrigation purposes. Bangladesh's agriculture including forestry, fisheries, and livestock as well as the settlement patterns are shaped by the availability and spatial and seasonal distribution of the country's water resources, both surface and ground. Differences in depth and duration of seasonal flooding on different soil and land types strongly influence the kinds of crops grown and crop rotations. Water is of course needed for household uses and is a crucial factor concerning sanitation and health. Other important uses of water relate to industry and navigation. Indeed,

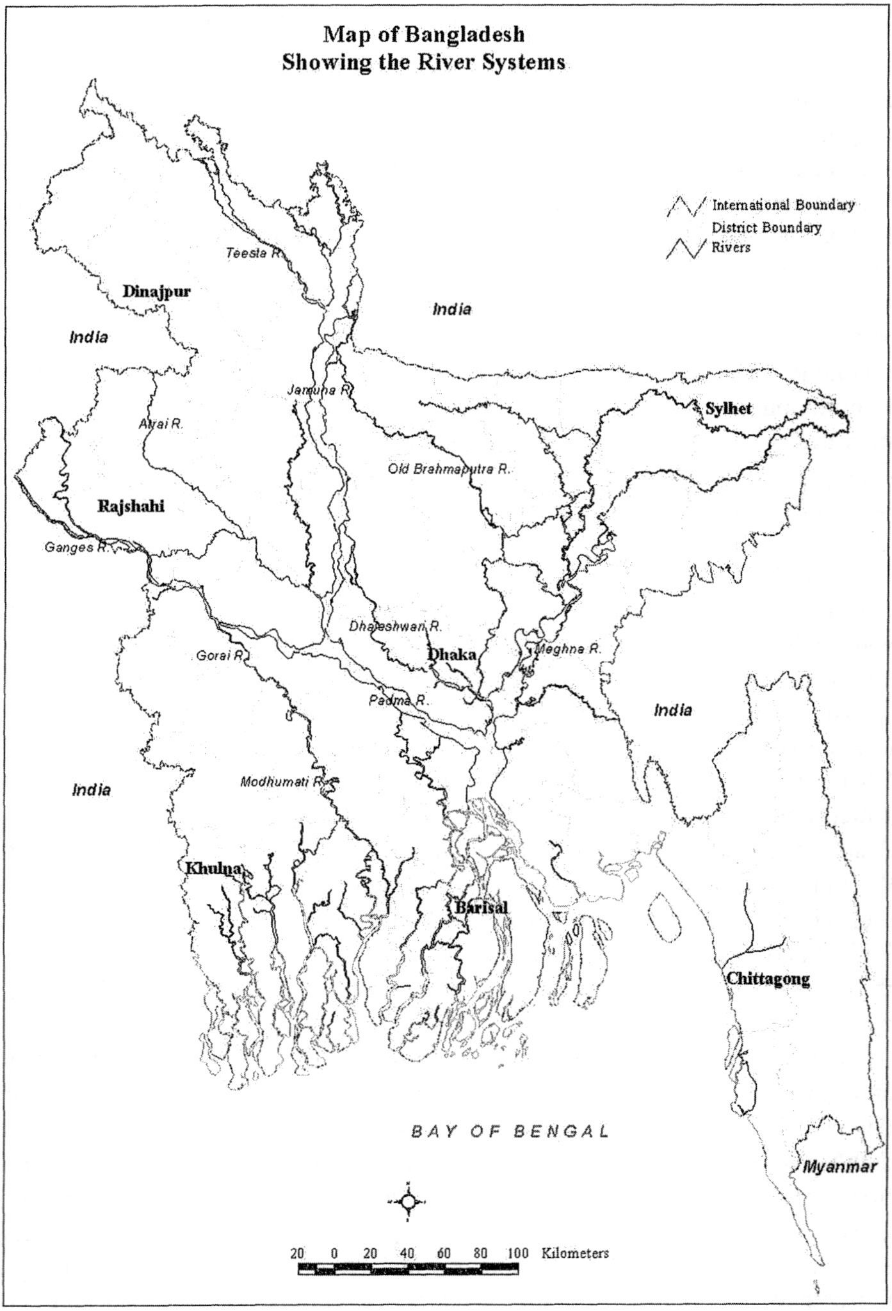

Figure 1. Map of Bangladesh highlighting the river systems.

Table I. Catchment areas of the GBM River Systems

Countries	Ganges		Brahmaputra		Meghna		GBM total	
	Area 1000 km^2	Share of the basin (%)	Area 1000 km^2	Share of the basin (%)	Area 1000 km^2	Share of the basin (%)	Area 1000 km^2	Share of the basin (%)
India	861	79	195	34	42	54	1,098	63
China	40	4	293	50	–	–	333	19
Nepal	140	13	–	–	–	–	140	8
Bhutan	–	–	45	8	–	–	45	3
Bangladesh	46	4	47	8	36	46	129	7
Total	1,087	100	580	100	78	100	1,745	100
Per cent of total	62		33		5		100	

Source: Adapted from Ahmad *et al.*, 1994.

inland water transport plays an important role in the movement of people and goods in Bangladesh.

Bangladesh is predominantly rural and agricultural and characterised by high population density, widespread poverty and unemployment, and low rates of economic growth. The majority of rural households depend on agriculture, fisheries, and other forms of primary activities. The socio-economic setting of the country is briefly described below.

Bangladesh has a very large population – 129.3 million as of January 2001, with a density of 876 persons per km^2, according to the preliminary findings of 2001 Population Census. The 1991 Population Census found a population growth rate of 2.17 per cent, which has declined to 1.5 per cent by 2001. About 77 per cent of the population lives in rural areas, and about 23 per cent in urban centres. (BBS, 2001). The estimated adult literacy rate as of 2000 is 64 per cent and the life expectancy at birth is estimated (1999/2000) at 61 years for both sexes (MoF, 2001). About 59 million people (45 per cent of the total population according to MoF 2001) are absolutely poor (in terms of average dietary intake of less than 2122 kcal), and about 30 million hardcore poor (consuming less than 1805 kcal per day).

Agriculture is still the mainstay of the economy; it generates about 25 per cent of the GDP and employs about 63 per cent of the labour force. Also, most of the other sectors/activities depends on agriculture (with the significant exception of ready-made garments) either for processing the products of or servicing the sector. Hence, agriculture remains the most important economic activity of the country, despite a recent decline in its relative contribution to GDP that was about 30 per cent in 1990/91, (MoF, 2001). During the second half of the 1990s, the annual average economic growth rate was 5.5 per cent, supported largely by agriculture which grew at an average annual rate of 4.4 per cent. Agriculture, however, critically depends upon weather patterns and hydrology, variables which are subject to variability due to climate change. Over the past 4 or 5 years, favourable weather patterns, farmers' determined efforts, and agricultural support programmes (input subsidies, agricultural credit, price support through purchase of foodgrains during or immediately after harvests and sale during agricultural lean seasons) of the government played important roles is sustaining such encouraging agricultural growth rates (Ahmad, 2000).

3. Floods in Bangladesh

Floods visit Bangladesh with monotonous regularity and studies have mainly been conducted going back many decades to ascertain the causes and dimensions of floods (See, for example, Mahalnobis, 1927; Mojumdar, 1941; Islam, 1980; Ahmad, 1989; Ahmad *et al.*, 2000; and Ahmed and Mirza, 2000). In general, there are four types of floods that occur in Bangladesh:

(i) floods due to inundation caused by the major rivers that carry large quantities of water entering Bangladesh from India in addition to in-country precipitation,
(ii) in-country rainfall induced floods,
(iii) flash floods in the eastern and north-central hill basins, and
(iv) floods caused by tidal storm surges.

Nearly 85 per cent of the in-country rainfall occurs during monsoon, June–October. This, together with drainage congestion in rivers, causes over-bank spills in many parts of the country. The floods become particularly severe and cause large scale damages when high levels of water flows from upper catchment areas join up with river flows generated by in-country rainfall. The high-intensity floods experienced by Bangladesh occurred when any two of the three major rivers reached peak flow conditions simultaneously. Over half of the country's landmass lies within 10 meters of the mean sea level and the flow at the sea mouth is retarded by a strong backwater effect. As a result, the discharge of floodwaters is constrained so that major floods become lengthy affairs, as was the case in 1998.

Previously, the area affected used to be taken as the measure of intensity of floods (Miah, 1988; World Bank, 1989). According to Miah (1988) the return period of normal floods is 2.25 years, of moderate to severe floods 4 years, of severe floods 7 years, and of catastrophic floods 33 to 50 years. The World Bank (1989) shows return periods of floods of different intensities with respect to the area covered (Table II). It was suggested that a flood like that of 1988 would occur once in about 100 years while one like that of 1987 in about 20 years. However, the interval between major (catastrophic) floods has been declining in recent times and is likely to do so even more in future as a result of flood accentuating consequences of climate change and sea level rise (see the introductory section). Note that floods inundating at least one-third of the country occurred on five occasions during the past half century: in 1955, 1974, 1987, 1988, and 1998, the interval having been 19 years, 13 years and 10 years successively.

Table II. Return period of floods characterized in relation to area covered

Return period	Affected area	Return period	Affected area
Years	% of the country	Years	% of the country
2	20	50	52
5	30	100	60
10	37	500	70
20	43		

Source: World Bank, 1989.

In accordance with the framework provided in Table II, floods of 1988 and 1998, both inundating about 60 per cent of the country, would be similarly categorized and would both be assigned a return period of about 100 years. But in reality, the 1998 flood event was far more destructive than the 1988 flood, as its duration (65 days) was several times of that (15~20 days) of the 1988 flood. (Choudhury, 2000). Now, the destructive capacity of flood is used to characterize its intensity. Obviously, when an area remains under flood water for a longer period, the greater the impact is likely to be. The flood intensity index is accordingly computed by multiplying the duration (number of days) of a flood in an area by the depth of the flood above the flood danger level for that area. Such calculations suggest that the 1998 flood was far worse than that of 1988 (Ahmed and Mirza, 2000).

Flood 1988 was the first recorded flood that lasted for such a long time. The reasons for the occurrence of such a catastrophic flood were many, which include: (a) excessive rainfall in the GBM catchment areas which was influenced by a La Nina phase of ENSO; (b) the peak discharge for all the three major rivers occurring at the same time; (c) a very strong backwater effect retarding drainage considerably; and (d) raised river bed levels due to continued sedimentation choking many small rivulets and reducing the drainage capacities of the major rivers leading to slower discharge of water. As a consequence of the 1998 flood about half the population of the country suffered losses of varying degrees in terms of crops, houses, and other property. Physical infrastructure were extensively devastated and disrupted. Relief needs were massive and rehabilitation needs were very demanding. Fortunately, loss of human lives was at an absolute minimum and the needs for relief and rehabilitation were met by highly organized and concerted efforts of the government, people in general, voluntary organizations, and others concerned. It was understood that losses incurred due to wide-scale damage of physical infrastructure, including settlements, could be minimized had the flood warning system be made more effective by way of increasing the lead-time for flood forecasting.

The latest flood that caused widespread suffering to the affected people as well as substantial losses of crop and property and damages to infrastructure occurred in the south-west Bangladesh in October 2000. The flood hit the area suddenly. Moreover, this part of Bangladesh did not experience a flood of this nature in the last 50 years or so. People were therefore not prepared to face it, particularly as it came upon them all on a sudden; there was no warning at all. In this case, floodwater predominantly came from the neighbouring Indian states of West Bengal and Bihar. A combination of excessive rainfall and over spillage and breaching of a number of reservoirs in India resulted in huge volumes of water in the rivers common to both the countries. No information about this was provided to Bangladesh. In fact, Bangladesh was caught unawares as waters, suddenly flowing into the country, caused a severe flood inundating several districts (including Meherpur, Chuadanga, Kushtia, Jhenidah, Jessore, and Satkhira). Millions of people were affected and excessive losses/damages were sustained by agriculture, fisheries, infrastructure, and trade and commerce. Also, there were extensive social and environmental adverse consequences. (Ahmad, E. *et al.*, 2001).

Had there been an effective data and information exchange arrangement between India and Bangladesh concerning sharing of flood-related data and information well in advance of a likely flood event, Bangladesh could act to minimize losses and sufferings.

4. Damages Caused by Floods in Bangladesh

All damages caused by a flood cannot be fully evaluated in monetary terms. However, Khan (1986) provided a conservative estimate of damages to crops and properties in monetary terms in the amount of US$ 200 million a year, on average. The damages caused by floods are of various types and nature and depends on a number of factors. Little or no damages are caused by the 'normal flood' with a return period of two years; people have learnt to live with such floods and developed their own response mechanisms. Moreover, it is often the case that post-flood crop production is facilitated by the organic matter enriched alluvium sediment carried by floodwaters (BCAS/PANOS, 1995). But crops grown in low-lying lands can be damaged, even destroyed by floods of moderate to severe intensity. Table III shows the estimated damages to foodgrains caused by floods for some selected recent years.

Table III. Damage to foodgrains in Bangladesh due to floods, selected recent years

Flood year	Damage	Flood year	Damage
	1000 metric tons		1000 metric tons
1963	62	1974	2000
1964	640	1976	1050
1968	1400	1980	500
1970	1900	1984	925
1971	1150	1987	800
1972	335	1988	2500
1973	1500	1998	3400

Source: Hofer and Messerli, 1994 for floods up to 1988 flood. For 1998 flood, authors' estimate based on relevant data obtained from Local Government Engineering Department (LGED) and other sources.

Not only foodgrains, but other crops are also damaged/destroyed by floods. Moreover, floods cause/aggravate the erosion of river banks, causing suffering to the affected people as not only their crops and movable properties are damaged, but they are also displaced becoming flood-refugees. Physical infrastructures such as roads, buildings, and factories are also damaged/destroyed by floodwaters. Academic years are shortened as roods to schools may be inundated/damaged, school buildings may be rendered unusable or used as relief camps, or children/families may be displaced by floodwaters. During floods, women and children tend to suffer the most because they are, generally, not as mobile and physically as resilient as men; women also stay back at home to keep their belongings safe. Even if they have food stuff, they may not be able to cook due to lack of dry biomass fuel; also, lack of access to safe drinking water is often a serious problem for them. Livestock die due to lack of shelter and fodder. All these damages and the necessary diverting of resources from development to relief activities constrain the developmental prospects as well. Obviously, to find ways and means of reducing adverse flood impacts is of crucial importance.

5. Flood Management Practices in Bangladesh

During the past four decades certain measures, mostly physical in nature, have been undertaken in Bangladesh towards combating floods. These interventions were made primarily to protect vulnerable croplands from inundation with a view to reducing crop losses, minimizing damages to properties and infrastructure, and saving lives.

5.1. STRUCTURAL MEASURES

The structural measures or engineering interventions undertaken include: (i) construction of embankments on river banks to prevent over-bank spillage; (ii) improving flow conditions in the channels and preventing riverbank erosions; and (iii) improvement of drainage conditions by various methods. Impounding of monsoon flows by constructing reservoirs to be released at controlled rates later on is a common flood mitigation measure used in GBM countries other than Bangladesh. Bangladesh has not tried this particular measure because of a combination of economic and a socio-physical reasons, including high rates of silt deposition, low topography, and risk of displacing many people.

The construction of embankments has been the most important structural flood mitigation measure implemented in Bangladesh. Embankments (adding up to over 8,300 km in length) have been constructed throughout the country since 1959 (Adhikari *et al.*, 2000). In addition, the Bangladesh Water Development Board (BWDB) constructed 1,695 flood control/regulating structures and 4,310 kms of drainage canals (BWDB, 1989). In most cases, the embankments were designed for protection against floods with a return period of twenty years. These embankments, therefore, provided little protection against high intensity floods of 1987, 1988 and 1998. In fact, the embankments themselves were damaged/breached in places.

The effectiveness of Flood Control and Drainage (FCD) and Flood Control, Drainage and Irrigation (FCDI) schemes came under scrutiny following the disastrous floods of 1987 and 1988 (IFCDR, 1992). From an analysis of a few of these projects it was found that there was some favourable effect on rice production during monsoon. But two country-wide evaluations found no significant increase in monsoon rice production on account of these projects (MPO, 1991; Khan, 1991). A five-year comprehensive review under the title of Flood Action Plan (FAP) for flood mitigation in the country was carried out by the Government of Bangladesh in the second half of the 1980s. Criticisms were later labelled against FAP, particularly from civil society, that it was donor driven (Boyce, 1990) and that it focused on structural measures and was of little practical usefulness. Nevertheless, it has generated a wealth of data and analysis and has also highlighted the usefulness of a few non-structural measures.

5.2. NON-STRUCTURAL MEASURES

Non-structural measures are employed to reduce the vulnerability of the flood affected people and improve their capacity to manage the consequences of floods (i.e., damages and losses). Such measures would include:

(i) Flood forecasting and warning services;
(ii) Flood-plain management measures such as flood-plain zoning and flood-proofing, including disaster preparedness and response planning;
(iii) Flood fighting including public health measures;
(iv) During and post-flood relief; and
(v) Flood insurance.

If flood warning is available to the people and concerned agencies a reasonable time in advance, steps can be taken to control damages and losses. Hence, in the context of responding to and managing the impact of floods, flood forecasting and warning is now considered to be one of the most important non-structural measures. While this measure is concerned with providing the lead time (i.e., time available between a forecast and the actual flooding), the other measures listed above constitute practical responses to be instituted towards improving flood management and reducing its adverse consequences.

6. Non-Structural Flood Management: Necessity for Regional Cooperation

Following Liberation of Bangladesh in December 1971, a Flood Forecasting and Warning Centre (FFWC) was set up in 1972 under the aegis of BWDB. But its capacity to provide flood warning well in advance remained very poor up to the end of the 1980s. Following the 1988 flood, Government of Bangladesh took an initiative to modernize the operation of the FFWC and enhance its institutional capacity. Currently the FFWC has the mandate to conduct the following activities: (i) collection of data on water level (in-country), (ii) forecasting meteorological conditions, (iii) forecasting floods, and (iv) dissemination of flood warning.

The FFWC collects data from 48 water level stations and 49 rainfall measuring stations around the country. Water levels are observed every three hours during the rainy season. The data recorded are transmitted to Dhaka through 61 wireless stations throughout the country. Data from another five stations outside the border of Bangladesh are provided by India. Based on these real-time data, mostly from within Bangladesh, validated computer models are applied to assess the floodability of particular vulnerable points and by using Geographical Information Systems (GIS) flood inundation maps are produced. These maps translate the extent, in percentage terms, of projected flooding for each administrative unit within the modeled area.

Based on the outputs of these exercises, various bulletins are issued giving forecasts of changes in river water levels at various flood prone locations. But, the lead time generated by the forecasting activity is 24 to 30 hours only. This is so because this exercise is primarily based on in-country real-time data from places which are very close to the locations to which the forecasts relate and (as explained later) limited and close-to-flooding time data from India; by the time these data are analysed and forecasts issued, floodwaters are nearly upon those vulnerable areas. Obviously, the responses can be improved significantly, if the lead time is increased by another 48–72 hours. And that can done only if real-time data are received from further upstream, i.e., from the co-riparians including India, Nepal, and Bhutan.

Given that 93 percent of the total GBM catchment area lies outside Bangladesh, flood forecasting and warning cannot but be an incomplete exercise unless there is meaningful and effective cooperation among the co-basin countries (Ahmad *et al.* 1994). Bangladesh undertook joint studies on floods bilaterally with India, Nepal, and Bhutan with a view to initiating meaningful regional cooperation in flood management. The studies (GoB/HMGoN, 1989; GoB/RgoB, 1989; GoB/GoI, 1990) have established the need for cooperation among the co- riparians, particularly for further development and strengthening of the flood forecasting and warning systems to face more effectively the challenges engendered by floods. It has also been recommended that in tandem with joint flood forecasting activities, necessary structural measures should be put in place.

7. Existing Regional Cooperation

Within the framework of bilateral cooperation between Bangladesh and India, Bangladesh receives water level data from five upstream stations in India, which are: Farakka on the Ganges, Dhubri and Goalpara on the Brahmaputra, Domohani on the Teesta, and Silchar on the Barak- Meghna. In addition, rainfall data are received from seven upstream stations in India, viz., Goalpara, Dhubri, Tura, Coochbehar, Siliguri, Jalpaiguri, and Agartala. It is very important for Bangladesh that it receives real-time data from all relevant locations in upper riparians as the country is the lowest riparian with respect to all three regional river (i.e., GBM) systems and drains most of their waters down to the Bay of Bengal. It, therefore, faces the brunt of the fury of flood caused by waters belting down the river systems. If real-time data with respect to all relevant locations in the upper riparians are available, starting from early stages as water levels begin to rise, Bangladesh can formulate and issue flood warnings well in advance to help the concerned people and agencies to undertake necessary mitigatory steps. But, not only that data relating to a limited number of locations are provided by India under the present arrangement, but also the transmission of data from India starts only when the hydro-meteorological conditions reach the warning stage, i.e., when water level has already risen to within one metre of the danger level, and rainfall exceeds 50 mm. From these data a lead time of only 24–30 hours can be generated for the central part of Bangladesh, and one not exceeding even four hours for areas near the borders.

Clearly, there is an urgent need in this context that cooperation with India is strengthened for obtaining more and timely relevant data. Cooperation with Nepal and Bhutan is also needed to acquire more real-time data from further upstream locations towards strengthening Bangladesh's flood forecasting and warning capability.

8. Strengthening of Regional Cooperation

The report (GoB/GoI,1990) of the Indo-Bangladesh Task Force on Flood Management, endorsed by both the governments identified more areas of cooperation to facilitate and support the preparation of forecasts which are more reliable and have longer lead times, thereby creating scope for improved flood preparedness. To this end, the report made the following proposals:

(i) Transmission of three-hourly real-time and daily forecast level data relating to the entire rainy season (May to October) starting from the early stages;

(ii) Inclusion in the data transmission arrangement of real-time and forecast data from further upstream stations such as Monghyr, Patna and Allahabad on the Ganges, Guwahati, Tejpur and Dibrugarh on the Brahmaputra, and Teesta Bazar, Gajaldoba, and Jalpaiguri on the Teesta;

(iii) Joint calibration of the hydrodynamic simulation model, being used in Bangladesh, by Bangladesh and India, which would improve the accuracy of forecasts and increase the lead times.

The joint studies with Nepal and Bhutan in 1989 have also recommended the exchange of hydrometeorological data on a broad scale between Bangladesh and Nepal and between Bangladesh and Bhutan (GoB/HMGoN, 1989; GoB/RgoB, 1989).

In order to improve the model framework for effective flood forecasting, Bangladesh has proposed exchange of several data sets with India in the following sectors (BANCID, 1997):

(i) River cross-section data at intervals of one section per 50 km up to Allahabad on the Ganges, Dibrugarh on the Brahmaputra, Silchar on the Barak-Meghna, and Teesta Bazar on the Teesta. The cross-section data may be updated before the monsoon every year.

(ii) Three-hourly water levels and daily forecasts from the following river stations:

River	Station	River	Station	River	Station	River	Station
Ganges	Allahabad	Brahmaputra	Dibrugarh	Barak/	Silchar	Teesta	Teesta bazar
	Patna		Tejpur	Meghna			Gajoldoba
	Monghyr		Guwahati				Domohani
	Farakka		Goalpara				Jalpaiguri
			Dhubri				

(iii) Daily discharge data from the following stations:
- on the Ganges at Allahabad, Patna, and Farakka;
- on the Kosi, Gandak, and Ghagra at their outfall;
- on the Brahmaputra at Dibrugarh, Tejpur, and Pancharatna;
- on the Barak-Meghna at Silchar; and
- on the Teesta at Gajaldoba and Domohani.

(iv) Daily rainfall data from the following stations:
- In the Ganges Basin: Monghyr, Gorakhpur, Champaran, Patna, Allahabad, Darbhanga.
- In the Brahmaputra Basin: Dhubri, Goalpara, Tura, Guwahati, Tejpur, and Dibrugrah.
- In the Barak/Meghna Basin: Karimganj, Amarpur, Agartala, Kailashar, Silchar, and Khowai town.
- In the Teesta, the Dharala and the Dudhkumar rivers: Jalpaiguri, Siliguri, Darjeeling, Kalimpong, Coochbehar.

(v) In medium and flashy rivers: water level, discharge, and rainfall data from a number of representative stations for hydrodynamic/rainfall-runoff modeling purposes of the following medium and flashy rivers:
- Mohananda
- Dharala (Jaldhaka)
- Dudhkumar (Torsa)
- Nitai
- Bhogai
- Someswari
- Kangsa
- Manu
- Khowai
- Dhalai
- Gumti
- Muhuri

Bangladesh and India both use similar technologies for observation and/or transmission of data, and similar methods for processing data for forecasting. Many of the hydrological stations in both countries have facilities for the observation of other parameters such as rainfall, humidity, and temperature, which are usually taken into account in the mathematical flood forecasting models. Both countries can, therefore, learn from each other's experiences and by sharing experiences and data, help enhance each other's capacity for making more accurate forecasts and for more effective issuance of warning.

For Bangladesh, the lowest riparian, comprehensive data-sharing arrangement with India for sure, but also with other GBM regional countries will be highly beneficial. Given that such a data sharing arrangement is in place, a state of the art integrated model framework can be developed towards enhancing flood preparedness in all affected areas across the GBM region, particularly in Bangladesh and the vulnerable states of India.

9. Concluding Remarks

Floods visit Bangladesh annually, monotonously. But the people have learnt to live with normal floods that inundate 23–30 per cent of the country's landmass. The interval between major (catastrophic) floods has been declining, while both the intensity and duration of such floods tend to be increasing. This outlook also generally applies to other GBM regional countries. Hence, more concerted flood management efforts are needed both in-country and regionally.

Thus, Bangladesh needs to further strengthen its institutional and technical capabilities providing accurate forecasts and warning and developing and launching, as needed, effective flood mitigation programmes in a coordinated fashion in which the people, the voluntary agencies, the government, and the other actors play their respective roles within a properly synchronized framework. At the same time, given the regional characteristics and implications of floods, efforts are needed to forge an effective GBM regional cooperation in flood management.

Bangladesh has been successful in cyclone forecasting and warning and in developing effective cyclone preparedness programmes and structures, which are very effective in saving lives and assets in coastal areas of Bangladesh. In the case of flood, Bangladesh can develop similarly effective flood forecasting and warning systems, given the real-time comprehensive data on water levels in rivers and rainfall are available from India in particular, but also from Nepal and Bhutan since well before danger levels are reached. A region-wide integrated approach based on comprehensive data sharing and joint modeling and scenario development efforts among the GBM countries should be highly useful in managing floods and mitigating its impacts in all the regional countries.

Human sufferings caused by disasters tend to rekindle human instincts to come in aid of fellow human beings who are in distress. Given that floods in the GBM region spell disastrous consequences for the various regional countries, joint action towards managing floods and minimizing their consequences may be found a worthwhile undertaking by all the countries. In other words, Disaster Diplomacy may provide the framework for collaborative efforts among the concerned countries (Kelman 2001). What has come to be known as Track II in South Asia, (viz. research, dialogue, identification and analyses of options available, and advocacy towards promotion of mutually beneficial regional cooperation conducted jointly by people outside the Track I, i.e., government, from different countries, including academics, experts, development practitioners, journalists, and others concerned) has been working over the past decade or so to promote regional collaborative efforts in the GBM region to solve common problems and implement activities in which each country has a stake for mutual benefit. (Ahmad *et al.*, 1994).

However, the major reasons behind a persisting regime of lack of cooperation in the GBM region include persisting fears and mistrust characterizing the attitudes of the regional countries towards one another, myopic views of governments, often coincidental with their own terms of office, and bureaucratic rigidities and hindrances. It has clearly emerged from a Track II study that GBM regional cooperation, particularly in the area of water development and management "offers to all countries gains far beyond anything that can be achieved by isolated national efforts" (Ahmad *et al.*, 1994, p.129). That is, cooperation is positive-sum, and win-win for all the participating countries. Indeed, collaborative flood management eminently qualifies as one such beneficial activity for all the GBM regional countries. Whether it is Disaster Diplomacy or Track II Diplomacy or any other framework that is adopted as the launching pad for promoting cooperation, what is crucially needed is enlightened and concerted efforts by all concerned to overcome the above mentioned underlying reasons hindering GBM regional cooperation so that the regional countries can undertake joint programmes/projects in flood management and other potential areas of cooperation.

References

Adhikari, K. D., Ahmad, Q. K., Malla, S. K., Pradhan, B. B., Rahman, K., Rangachari, R., Rasheed, K. B. S., and Verghese, B. G (eds.): 2000, *Cooperation on Eastern Himalayan Rivers: Opportunities and Challenges*, Bangladesh Unnayan Parishad (BUP), Dhaka/Konark Publishers Pvt. Ltd., New Delhi, India.

Ahmad, E., Rahman, S. M. M., Hossain, N. H. A., and Rahman, K.: 2001, Flood 2000: Causes and consequences. Presented to a seminar on *Flood 2000 in the South Western Region of Bangladesh*, organized by Bangladesh Unnayan Parishad (BUP), Dhaka, January 2001.

Ahmad, M.: 1989, *Flood in Bangladesh*, Community Development Library, Dhaka, Bangladesh.

Ahmad, Q. K.: 2000, Bangladesh's development strategy and the role of external assistance, German assistance in particular, in *Development Cooperation at the Dawn of the Twenty First Century: Bangladesh-German Partnership in Perspective*, Bangladesh Institute of International and Strategic Studies (BIISS), Dhaka, Bangladesh.

Ahmad, Q. K., Biswas, A. K., Rangachari, R., and Sainju, M. M. (eds.): 2001, *Ganges-Brahmaputra-Meghna Region: A Framework for Sustainable Development*, University Press Limited, Dhaka, Bangladesh.

Ahmad, Q. K., Chowdhury, A. K. A., Imam, S. H., and Sarker, M. (eds.): 2000, *Perspectives on Flood 1998*, University Press Ltd., Dhaka, Bangladesh.

Ahmad, Q. K., Verghese, B. G., Iyer, R. R., Pradhan, B. B., and Malla, S. K. (eds.): 1994, *Converting Water into Wealth: Regional Cooperation in Harnessing the Eastern Himalayan Rivers*, BUP/Academic Publishers, Dhaka, Bangladesh.

Ahmed, A. U. and Mirza, M. M. Q.: 2000, Review of Causes and Dimensions of Floods With Particular Reference to Flood '98: National Perspectives, In: Q. K. Ahmad, A. K. A. Chowdhury,

S. H. Imam and M. Sarker (eds), *Perspectives on Flood 1998*, University Press Ltd., Dhaka, Bangladesh.

BANCID: 1997, *Proceedings of Seminar on Evolution of a Scientific System of Flood Forecasting and Warning in the Ganges, Brahmaputra and Meghna River Basins*, organized by Bangladesh National Committee of the International Commission on Irrigation and Drainage, Dhaka, Bangladesh.

BBS: 2001, *Population Census 2001, Preliminary Report*, Bangladesh Bureau of Statistics (BBS), GoB, August 2001, Dhaka, Bangladesh.

BCAS/PANOS: 1995, *Rivers of Life*, Bangladesh Centre for Advanced Studies (BCAS) and PANOS International, Dhaka, Bangladesh.

Boyce, B.: 1990, Birth of a megaproject: Political economy of flood control in Bangladesh, *Environmental Management* **14**(4), 419–428.

BWDB: 1989, *Study on the Causes and Consequences of Natural Disasters and Protection and Preservation of the Environment*, Bangladesh Country Report, Government of Bangladesh, Dhaka, Bangladesh.

Choudhury, A. M.: 2000, Flood 1998: Oceanic perspective, In: Q. K. Ahmad, A. K. A. Chowdhury, S. H. Imam and M. Sarker (eds), *Perspectives on Flood 1998*, University Press Ltd., Dhaka, Bangladesh.

GoB/GoI: 1990, *Report: Indo-Bangladesh Task Force on Flood Management*, Government of Bangladesh (GoB) and Government of India (GoI), Dhaka and New Delhi.

GoB/HMGoN: 1989, *Report on Flood Mitigation Measures and Multipurpose Use of Water Resources*, Government of Bangladesh and His Majesty's Government of Nepal (HMGoN), Dhaka and Katmandu.

GoB/RgoB: 1989, *Report on Flood Control and Flood Mitigation*, Government of Bangladesh and Royal Government of Bhutan (RGoB), Dhaka and Thimpu.

Hofer, T. and Messerli, B.: 1994, *Floods in Bangladesh: Process Understanding and Development Strategies*, Institute of Geography, Geneva, Switzerland.

IFCDR: 1992, *Pilot Programme to Improve Management of Flood Control, Drainage, and Irrigation Projects*, Institute of Flood Control and Drainage Research (IFCDR), Bangladesh University of Engineering and Technology (BUET), Dhaka, Bangladesh.

IPCC: 2001, *Climate Change 2001: Impacts, Adaptation, and Vulnerability – Summary for Policymakers and Technical Summary of the Working Group II Report*, Third Assessment Report of the Intergovernmental Panel on Climate Change (IPCC), Cambridge University Press, U.K.

Islam, M. A.: 1980, Floods in Bangladesh: Causes, consequences and adjustments. In: *Proceedings of International Conference on Bangladesh Floods: What Are the Solutions*, Montreal, Canada, May 11–13, jointly organized by Concordia University and International Society for Bangladesh.

Kelman, I.: 2001, Implementing and applying disaster diplomacy (http://www.disasterdiplomacy.org).

Khan, H. R.: 1991, Impact of flood control and drainage projects on agricultural production in Bangladesh. Presented to a seminar organized jointly by Institution of Engineers and ASCE (Bangladesh IG), Dhaka, 6 July 1991.

Khan, M. A. H.: 1986, Floods: Problems and prospects – A regional review. In: *Proceedings of the Regional Conference on Floods and Erosion*, Institution of Engineers, Dhaka, Bangladesh.

Mahalnobis, P. C.: 1927, *Report on Rainfall and Floods in the North Bengal 1870–1922*, Irrigation Department, Government of Bengal, Bengal Secretariat (Book Depot), Calcutta, India.

Miah, M. M.: 1988, *Flood in Bangladesh: A Hydromorphological Study of the 1987 Flood*, Academic Publishers, Dhaka, Bangladesh.

MoF: 2001, *Bangladesh Economic Review 2001*, Ministry of Finance (MoF), GoB, June 2001, Dhaka.

Mojumdar, S. C.: 1941, *Rivers of the Bengal Delta*, Bengal Government Press, Alipore, Bengal.

MPO: 1991, *National Water Plan*, Master Plan Organization (MPO) for Ministry of Irrigation, Water Development and Flood Control, GoB, Dhaka, Bangladesh.

Rahman, M. R. and Chowdhury, J. U.: 1998, Impacts of flood control projects in Bangladesh. In: M. M. Ali, M. M. Hoque, R. Rahman and S. Rashid (eds), *Bangladesh Floods – Views from Home and Abroad*, University Press Limited, Dhaka, Bangladesh.

Rasheed, K. B. S.: 2001, Water resources in Bangladesh: Development and management perspectives. Paper presented at the Regional Conference on Water Resource Management in South Asia, organized by the U.S. Education Foundation in India, Hyderabad, India, 6–7 September.

World Bank: 1989, *Bangladesh Action Plan for Flood Control*, South Asia Department, Washington, D.C.

Natural Hazards **28:** 199–210, 2003.

Economics of Flood Protection in India

SUJATA GUPTA, AKRAM JAVED and DIVYA DATT
Tata Energy Research Institute, (Teri) Darbari Seth Block, Habitat Place, Lodhi Road, New Delhi – 110 003, India

(Received: 9 July 2001; accepted: 4 March 2002)

Abstract. The peculiar rainfall pattern in India renders the country highly vulnerable to floods. Forty million hectares of land, roughly one-eighth of the country's geographical area, is prone to floods. Each year, floods cause extensive damage to life and property, losses being exacerbated by rapid population growth, unplanned development and unchecked environmental degradation. The country has been tackling the problem through structural and non-structural measures. While non-structural measures like flood forecasting aim at improving the preparedness to floods by seeking to keep people away from floodwaters, structural measures involve the construction of physical structures like embankments, dams, drainage channels, and reservoirs that prevent floodwaters from reaching potential damage centres. Almost 48% of the vulnerable area has been provided with reasonable protection, though floods continue to cause widespread losses year after year. This paper examines the incidence of floods and the trends in consequent losses in the eastern region of the country – one of the most vulnerable – with the objective of studying the efficacy of flood protection measures in the region. Based on a simple regression exercise for three highly vulnerable states in the region, the paper argues that flood protection measures have been inadequate in controlling losses and reducing vulnerability. Regressions for the three states over the period 1971 to 1996 indicate that the level of protection is an insignificant explanatory variable in explaining the number of people (adjusted for increases in density) affected by floods; while area affected, as an indicator of the intensity of floods remains the main loss-determining factor.

Key words: Natural disasters, Flood Management, Eastern India, Population growth, Geographical Information System (GIS)

1. Introduction

During the last 30 years, natural disasters have caused the death of at least three million people worldwide and have affected many more millions. The economic costs of natural disasters are on an alarming rise: compared to the 60s, the economic loss burden due to disasters has increased eight fold, discounting inflation (Sinha and Sharma, 1999). In the last ten years alone, major natural disasters have caused economic losses of more than US$ 400 billion. With such huge losses to life and property, natural disasters have become a serious threat to development especially in developing countries, already pressed for resources. Ninety percent of natural disasters and ninety five percent of all deaths from natural disasters occur in developing countries. Among the top ten countries, in terms of average annual loss to life due to natural disasters, all except Australia are developing countries.

China tops the list with average annual population affected being about 99 million and India follows second with 56.6 million.[1]

Apart from natural forces, human activity can significantly impact the incidence of natural disasters. Environmental degradation – which is often a result of economic development and associated human settlement patterns that ignore appropriate resource management – can increase a country's vulnerability to natural hazards and exacerbate the impacts. For instance, excessive logging and land clearance for cultivation increases erosion of the soil and can induce flash floods. This in turn causes siltation in river courses, intensifying floods.

Floods continue to be one of the most severe natural disasters, particularly in Asia. The intensity of flood disasters appears to have increased in the region during the past decade. Findings of a regional survey in 1998 by the United Nations Economic and Social Commission for Asia and the Pacific (ESCAP)/International Decade for Natural Disaster Reduction Secretariat put countries in the Asian region as most affected by annual floods, especially during the past few years (Le Huu Ti, 1999).

India is one of the most flood prone countries in the world. Twenty-three out of thirty two-states/union territories in the country are subject to floods and 40 million hectares of land, roughly one-eighth of the country's geographical area, is prone to floods. The National Flood Control Program was launched in the country in 1954. Since then sizeable progress has been made in the flood protection measures. By 1976, nearly one third of the flood prone area had been afforded reasonable protection; considerable experience has been gained in planning, implementation and performance of flood warning, protection and control measures.

This paper examines the incidence of floods in the country along with trends in related losses with the objective of studying the efficacy of flood protection measures in the country. After a background on the nature and extent of the flood problem in the country, the paper examines the main flood management measures common in India- both structural and non-structural. The paper then attempts to study the role of such measures in controlling floods in the eastern region of the country – one of the most flood prone. Due to limited district level data on flood-related losses, the analysis is based mainly on statewide data.

2. Floods: A Major Natural Disaster in India

India's unique sub-continental dimensions, geographical position and the behaviour of the monsoons make the country one of the most disaster prone in the world. The Indian sub-continent is highly vulnerable to droughts, floods, cyclones and earthquakes, though landslides, avalanche and bush fires too, frequently occur

[1] Top ten countries in terms of average annual number of people affected in 1987/96 are China (99.07), India (56.56), Bangladesh (18.57), Ethiopia (4.02), Philippines (3.69), Australia (2.28), Thailand (1.67), Sudan (1.48), Malawi (1.44), and Pakistan (1.40). Figures in brackets indicate in million average number of people affected annually (Sinha and Sharma, 1999).

in the Himalayan region of the country. Of these, the most devastating ones are floods, cyclones and earthquakes. The country is affected by these disasters year after year in one part or the other.

India is the worst flood-affected country in the world after Bangladesh and accounts for one-fifth of the global death count due to floods (Anon, 1991). The average annual rainfall in India is about 120 cm, slightly more than the global mean of 99 cm. Floods are caused mainly due to the extreme spatial and temporal variation in the rainfall pattern in the country – 75% of the total annual rainfall is concentrated over the short monsoon period of three to four months, resulting in a very heavy discharge from rivers during this period. The most flood prone area in India lies in the Ganga-Brahmputra-Meghana river basin, which carries 60% of the nation's total river flow.

An estimated 8 million hectares of land are affected annually; the cropped area affected annually ranges from 3.5 million ha during normal floods to 10 million ha during worst floods. The chronic flood prone basins are Ganga and Brahmaputra covering northern and north-eastern parts of the country. The five most flood prone states are Uttar Pradesh (U.P.), Bihar, West Bengal (WB), Assam and Orissa. Flooding cause wide spread damage to human lives, crop, plantation, property, infrastructure and business. Floods also leave in their wake epidemiological threats, breakdown of social order and temporary/permanent migration. Rapid population growth and unplanned development is accelerating vulnerability to floods as deforestation worsens the incidence of floods and settlements encroach into flood-prone lands. Roughly 30 million people in the country are affected by floods and more than 1500 lives lost each year (Figure 1).

3. Flood Management in India

The general approach to tackle the problem of floods has been in the form of structural and non-structural measures. Structural measures involve the construction of physical structures that prevent flood waters from reaching potential damage centres. Out of the 40 Mha area that is flood prone, roughly 32 Mha is protectable: of this about 59.6% has been provided with reasonable protection (Ninth Five Year Plan). Non-structural measures like flood forecasting on the other hand aim at improving the preparedness to floods by seeking to keep people away from floodwaters. The various structural and non-structural measures adopted in the country are briefly discussed below.

3.1. STRUCTURAL MEASURES

Structural measures have an ancient origin: private embankments have existed in the country for centuries. The inadequacy of individual efforts in the area of flood control led to government interest in the problem and a number of well-planned embankments were constructed beginning the 1950s in commanded areas of canal

Figure 1. Human lives lost due to floods in India. Source: Sinha and Sharma, 1999.

systems in northern India and the deltaic tracts of east flowing rivers in the South. The major thrust of flood protection programmes undertaken in the country so far has been in the areas of embankments, drainage channels, dams and reservoirs, drainage and afforestation.

3.1.1. *Embankments and Flood Walls*

The principal purpose of embankments and flood walls is to confine flood waters to the stream channel and a selected portion of the flood plain. These barriers protect only the land area immediately lying behind them and are effective only against flood depth up to a certain predefined level for which they were designed. Floods exceeding the designed levels can cause disastrous loss of life and property. Embankments remain the most common and economical form of protection to provide immediate relief from inundation. It is estimated that by 1997, about 16,200 km length of embankments had been constructed across the country however the maintenance of embankments though has been poor due to paucity of funds. Lack of adequate maintenance can lead to the disruption of the structures in periods of heavy rains, causing large-scale devastation in settlements in the plains. It is thus necessary to ensure proper maintenance of embankments involving the beneficiaries. It is also necessary to educate the masses about the consequences of occupying flood plains. Since the flood plain is a natural right of the river, any kind of permanent structures should not be allowed within the flood plain of major rivers. This requires extensive flood plain zoning of major rivers demarcating their extent.

Besides, people living in flood prone areas should be given appropriate training in flood fighting, disaster mitigation and management and relief operations.

3.1.2. *Dams and Reservoirs*

Several dams and reservoirs in the country like the Ukai reservoir, the Bhakra and Beas Dams and the Hirakud Dam are shining examples of flood control. Apart from moderating floods, reservoirs are generally multipurpose projects, serving other uses such as irrigation, power generation, meeting industrial and domestic water requirements. There are however, several social and environmental costs associated with large dams. These include evacuation and resettlement of people in reservoir area, inundation of forests and changes in downstream hydrology. These impacts should be thoroughly addressed at the project investigation stage and compensation packages to address socio-economic disruptions should be designed in consultation with the affected people.

3.1.3. *Channel Improvement*

The flood carrying capacity of a river channel can be improved by deepening and widening it and by removing sharp bends etc. Action on these lines has been taken on the Jhelum in the Kashmir Valley.

3.1.4. *Drainage Improvement*

Inadequate artificial or natural drainage can adversely affect soil fertility through water logging, as has happened in Punjab and Haryana. A number of drainage schemes have been taken up in these States and others including Uttar Pradesh and West Bengal.

3.1.5. *Diversion of Flood Waters*

Diverting a part of the flood from a river into an adjacent valley is an another flood management option. This has been successfully tried in Andhra Pradesh and Rajasthan. In Rajasthan, about 340 cumecs (cubic metres per second) of the flow of the Ghaghar, which was flooding large irrigated areas, has been diverted to a number of depressions in sand dunes south of the irrigated belt using a 51 km long diversion channel.

3.2. NON-STRUCTURAL MEASURES

It has been realized over time that structural measures can only partly contain damages. Flood plains belong to the river and must be used judiciously, even vacated when the situation demands. Non-structural measures aim at disaster preparedness and reducing the loss burden from disasters. These measures include the following.

3.2.1. *Flood Plain Management and Zoning*

Heavy encroachment of flood plains has been responsible for the increasing trend of damage over time. The basic principle of flood plain management is to regulate the land use in flood plains in order to minimise the damage due to floods while deriving maximum benefits from these. This is done by determining the locations and the extent of areas likely to be affected by floods of different intensities/frequencies and to develop these areas accordingly, through zoning regulations such that damages from floods are minimised. However flood plain zoning has yet to taken up in a big way in most states in India.

3.2.2. *Flood Proofing*

Flood proofing measures help greatly in mitigating distress and providing immediate relief to the population in flood prone areas. Flood proofing is a combination of structural measures and emergency action. It usually entails providing raised platforms for flood shelter for people and cattle. Public utilities are generally raised above flood levels in order to make basic facilities "flood proof".

3.2.3. *Forecasting, Monitoring and Early Warning Systems*

An advanced system of forecasting, monitoring and issuing early warnings plays a central role in determining whether a flood will assume disastrous proportions. With timely information, loss of life and movable property can be prevented substantially. Forecasting is also essential for efficient functioning of reservoirs.

The Central Water Commission (CWC), Ministry of Water Resources has a network of 132 forecasting stations covering most interstate flood prone rivers besides inflow forecasts for 25 major reservoirs of the country. CWC maintains close liaison with the local civil authorities to communicate advance warning for appropriate prevention/mitigation measures. On an average, 6000 flood forecasts are issued every year.

The other non-structural measures relate to modifying the loss burden through disaster relief, flood fighting measures and flood insurance.

3.2.4. *Flood Disaster Relief*

The primary responsibility for rescue, relief and rehabilitation measures lies with the state government. The role of the central government is supportive, in terms of physical and financial resources. Financial resources are made available from the CRF (Calamity Relief Fund) and the NFCR (National Fund for Calamity Relief). The CRF exists for each state with the allocated amount contributed by the Central and State Governments in the ratio 3 : 1 as per the recommendations of the Tenth Finance Commission. The NFCR has been created to deal with calamities of rare severity and is managed by a National Calamity Relief Committee (NCRC) under the Ministry of Agriculture (MoA). The Natural Disaster Management Division within the MoA, along with the National Centre for Disaster Management (NCDM) within the Indian Institute of Public Administration (IIPA), New Delhi are the nodal agencies for natural disaster management in the country.

Apart from these initiatives, there are various programmes being implemented by various Ministries/Departments that impact flood management. These include Drought Prone Area Programme, Desert Development Programme and Technology Mission on Drinking Water by the Ministry of Rural Areas and Employment; National Watershed Development Programme for Rainfed Areas, Soil Conservation in the Catchments of River Valley Projects and Catchments of Flood Prone Rivers by the MoA; and Flood Control Scheme by the Ministry of Water Resources.

3.2.5. *Flood Fighting*

Flood fighting measures are undertaken immediately after a flood, with the objective of minimising losses and human misery. Measure include evacuation of flood victims, provision of food and other amenities, including first aid, plugging of breaches, restoration of road/rail links and public assets like water supply, power etc and voluntary initiatives by organisations like the Red-Cross, local people etc. Local people can play a key role in minimizing losses due to any disaster and appropriate training must be imparted to local communities, NGOs and voluntary organizations.

4. Economics of Flood Protection: An Analysis

Expenditure on flood management has increased significantly over the last few plan periods; the amount however remains a small share of total outlays in the irrigation sector, averaging about 6%. The inadequacy of flood management expenditure comes into focus when the amount is compared with the expenditure on relief and rescue operations. Table I compares expenditure on flood management vis-à-vis expenditure on actual relief operations for a few years in the 1990s.

Inadequacy of flood management in the country is apparent from the large area that is prone to floods but without a reasonable level of protection. Figure 2 shows affected area in the country and how much of that is reasonably protected. The spatial representation of the district level data was analyzed through Geographical Information System (GIS). The districts were classified into three categories i.e., (i) those having less than 25% of the geographical area threatened by floods, (ii) those having 25–50% of the geographical area threatened by floods and (iii) those having more than 50% of the geographical area threatened by floods. Of these districts, those with a reasonable level of protection are shown by dots. For the purpose of this study, reasonable level of protection was defined as being greater than or equal to the national average of 48% (i.e., for the country as a whole, 48% of the area threatened by floods is reasonably protected). The same information is provided in Table II, which shows for each state the number of districts that are vulnerable to floods and of these the number that have a *reasonable* degree of protection. The

Table I. Expenditure flood protection vs flood relief (Rupees billion)

Year	Expenditure on flood control	Expenditure on relief
1990–91	1.97	6.46
1991–92	2.64	5.91
1992–93	3.30	8.93
1993–94	3.66	5.64
1994–95	3.08	N.A
1995–96	3.12*	7.29
1996–97	3.07*	N.A

* Provisional; N.A: Not available. Source: Central Water Commission, GoI (Government of India).

Table II. Number of districts threatened and not reasonably protected*

State	District area prone to floods		
	0–25%	25–50%	>50%
Andhra Pradesh	10 (8)	3(1)	0
Assam	14 (7)	8 (3)	0
Bihar	6 (5)	7 (4)	12 (9)
Gujarat	9 (7)	3 (3)	0
Haryana	4 (2)	8 (2)	2 (–)
Kerala	9 (7)	4 (–)	0
Punjab	0	2 (–)	10 (5)
Orissa	3 (1)	1 (1)	0
Uttar Pradesh	20 (20)	25 (23)	9 (7)
West Bengal	2 (1)	5 (4)	7 (5)

*Figures in parenthesis indicate number of districts with protected area less than the national average of 48%. Source. Ministry of Urban Development, 1997.

states of Punjab, Bihar, Uttar Pradesh and West Bengal have a high number of districts that have a large area prone to floods. As can be inferred from Figure 2 and Table II, the eastern region, specifically the states of Bihar, Uttar Pradesh and West Bengal, is one of the most vulnerable and least protected.

That the increase in area protected (Table III) for the three most vulnerable states being studied – Bihar, Uttar Pradesh and West Bengal – has been inadequate is reflected in the non-decreasing trend of annual losses per unit area affected. Table

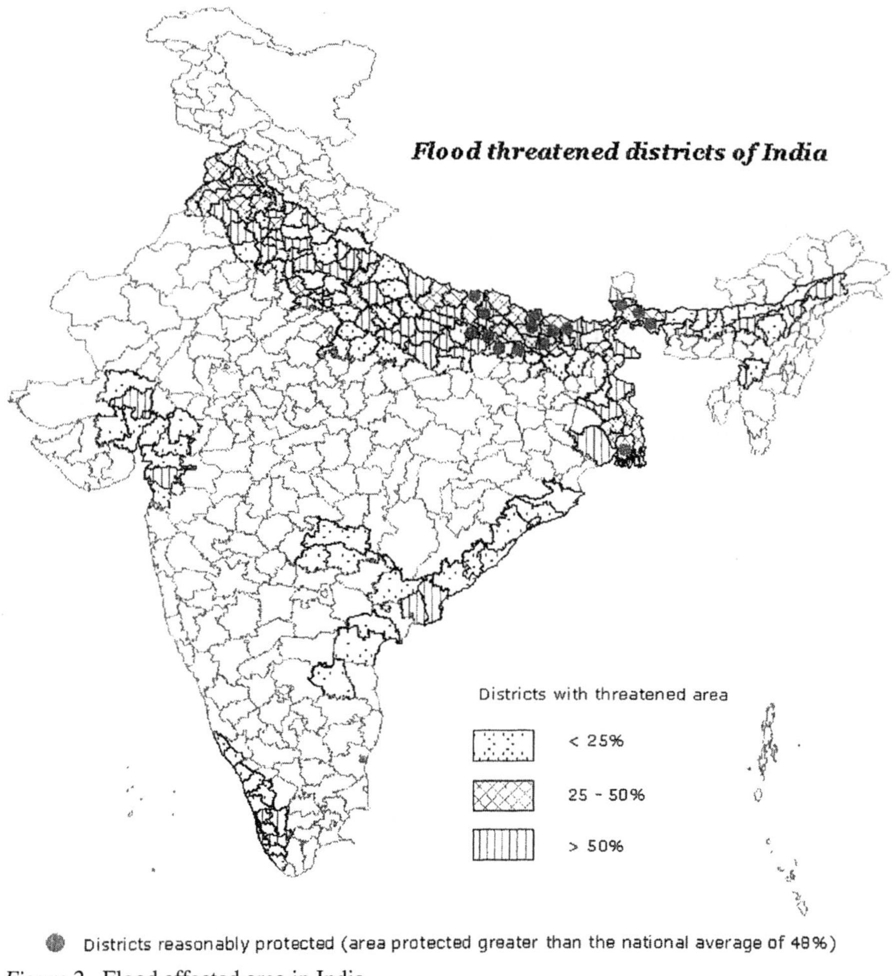

Figure 2. Flood affected area in India.

IV shows the trends in the three states for population affected per unit area affected by floods. Similar trends can be observed for human lives lost and cattle lost.[2] It is clear that the rate of growth of population and the increase in its density have outstripped the pace of flood management in these States.

In general, the population affected by floods should be expected to depend on the area affected by floods (an indicator of the intensity of the flood) and the level of protection. As the level of protection and disaster management increase, the number of people affected should decrease. A simple regression exercise was undertaken to study these relationships. Regressions for the three states over the period 1971 to 1996 indicate that the rise in the level of protection (Table III) is

[2] There are other indices of losses such as value of loss in crops and property, however the analysis was restricted to non-monetised data to simplify comparison.

an insignificant variable in explaining the number of people (adjusted for increases in density) affected by floods,[3] with area affected, as an indicator of the intensity of floods being the dominant determinant (Table V). In other words, despite more flood prone area being protected, people in affected areas are still highly vulnerable to floods. Inter state comparison also points to a similar conclusion. West Bengal with a high degree of protection, (33% of the vulnerable area being protected in West Bengal and Bihar as against 11.5% in UP), is most severely prone to floods as indicated by the large number of people and houses affected and lives lost per unit area affected. One factor that may explain this is the high density of population in the state (767 persons per km as against 496 in Bihar and 472 in UP as per the 1991 census). These results thus indicate that even though the extent of protection has increased overtime, it has not kept pace with demographic trends and has been inadequate in reducing the vulnerability to floods.

Table III. Area prone to floods and reasonably protected (million hectares)[4]

	Uttar Pradesh	Bihar	West Bengal
Area liable to floods adopted by R.B.A.*	7.34	4.26	2.65
Area reasonably protected upto			
1978	0.74	1.57	1
1987	1.42	1.87	1.75
1993–94	1.54	2.93	2.08

* Rashtriya Barh Aayog, (National Commission on floods), GoI, 1980. Source. Based on data from Central water Commission, GoI.

Table IV. Number of people affected per unit area affected – decadal averages

Period	Uttar Pradesh	Bihar	West Bengal
1961–1970	3.25	4.02	2.43
1971–1980	4.04	4.53	4.54
1981–1992	4.85	5.28	6.17

Data for the following years was not available: 1961, 1989, and 1990.
Source: Based on data from Central Water Commission, GoI.

[3] Annual densities and area protected for the period 1971 and 1994 was estimated by interpolating data for the available years.

[4] Here area reasonably protected is as defines by the government.

Table V. Population affected (adjusted for increases in density) as a function of area protected and area affected – Regression results

S. No.	State	Independent variables		Regression
		Area protected	Area affected	coefficient
1	Uttar Pradesh	Negative and not significant	Positive and significant	0.84
2	Bihar	Negative and not significant	Positive and significant	0.93
3	West Bengal	Negative and not significant	Positive and significant	0.85

5. Way Forward

There is clearly an inadequate level of protection in the country against floods. Though non-structural measures improve the preparedness to floods and reduce losses, the necessity of structural measures would always remain to reduce the extent of physical damage caused by floods. In this context it needs to be highlighted that a number of flood management schemes are not managed well- construction activities lag behind schedule and spill over from one period to the next. Inordinate delays lead to cost overruns, besides putting off the benefits to be realized from these schemes. It is thus necessary that resources are utilized efficiently. Apart from timely execution of new projects, it is as important to pay adequate attention to the maintenance of embankments and drainage channel. Poor maintenance is noted as a major weakness in the flood control system of the country. Maintenance of embankments, drainage channels, soil conservation and watershed management should be the responsibility of the beneficiaries, hence involvement of the community in the operation and maintenance of flood control measures through appropriate training and capacity building is necessary. People's participation in flood control through community management of flood protection structures and training in emergency action would go a long way in improving the preparedness to floods and reducing the damages caused by this natural phenomenon. On the other hand, there is enough evidence to suggest that exploitation of the environment can significantly impact the frequency and severity of natural disasters like floods. The loss of natural vegetation, particularly forests is a major cause of preventable natural disasters like floods. Thus along with investment in flood control, the government should ensure that environmental degradation is arrested and unplanned development in riverbanks checked through appropriate policies and awareness programmes. At the national level, flood control and management planning need to be integrated into development planning for the country.

References

Anon: 1991, *Floods, Flood Plains and Environmental Myths: State of India's Environment – A Citizens' Report*, Center for Science & Environment, New Delhi.

Central Water Commission: 1989, *Water and Related Statistics*, CWC (Statistics Directorate), New Delhi.

Gupta, A.: 2000, Information Technology and Natural Disaster Management in India, *GIS Development* (http://www.gisdevelopment.net/aars/acrs/2000/ts8/hami0001.shtml)

Le, H. T.: 2000, Experiences of flood control and management in Asia relevant to economic and social development strategies for central Vietnam in the twenty first century, Paper presented at the "International Conference on socio-ecomic strategies in central Vietnam at the dawn of the twenty first century". University of Da Nang in cooperation with USA–Viet Nam Foundation, Da Nang, 20–21 April 2000 http://asia.anu.edu.au/mekong/flood.html

Ministry of Urban Development: 1997, *Vulnerability Atlas of India* (Earthquake, windstorm and flood hazard maps and damage risk to housing), Building Materials & Technology Promotion Council, Government of India.

Rashtriya Barh Ayog: 1980, Report – Volume 1, Ministry of Energy and Irrigation (Department of Irrigation), Government of India, New Delhi.

Sinha, A. and Sharma, V. K.: 1999, *Culture of Prevention, Natural Disaster Management: India*, National Centre for Disaster Management, Indian Institute of Public Administration Natural Disaster Management Division, Department of Agriculture & Cooperation, Ministry of Agriculture, New Delhi.

Planning Commission: 1992, *Eighth Five Year Plan: 1992–1997 Vol. 2*, Planning Commission, New Delhi.

Planning Commission: 1997, *Ninth Five Year Plan: 1997–2002 Vol. 2*: Thematic Issues and Sectoral Programmes, Planning Commission, New Delhi.

9 781402 010682

GPSR Compliance

The European Union's (EU) General Product Safety Regulation (GPSR) is a set of rules that requires consumer products to be safe and our obligations to ensure this.

If you have any concerns about our products, you can contact us on ProductSafety@springernature.com

In case Publisher is established outside the EU, the EU authorized representative is:

Springer Nature Customer Service Center GmbH
Europaplatz 3
69115 Heidelberg, Germany

Batch number: 10614746

Printed by Printforce, the Netherlands